I0011731

Solve it with PYTHON!

Solve it with PYTHON!

A PROGRAMMING GUIDE TO EASE YOUR SCIENCE
AND ENGINEERING CHALLENGES

Javier Riverola Gurruchaga

Solve it with PYTHON!
by Javier Riverola Gurruchaga

Copyright © 2019 by Javier Riverola Gurruchaga. All rigths reserved.

Independently published in the United States of America by Amazon KDP.

Front cover ilustration: The Flammarion engraving (Camille Flammarion's 1888 book L'atmosphère: météorologie populaire) coloured by Houston Physicist, available under Creative Commons Attribution-Share Alike 4.0 International license.

ISBN: 9781689604109

No part of this book may be reproduced in any form, without written permission from the author, except for the use of brief quotations in a book review.

To my mother and my wife, the women of my life.

Contents

Preface

This book is intended for students and professionals in science and engineering who face problems that frequently arise in the mathematical arena. Instead of going to generic recipes, I have preferred to be practical and illustrate the programming techniques with simple examples that go straight to the point, without going into a detailed explanation of each step. Besides, I have given up impressing the reader with the superpower and sophistication that Python is capable of reaching in favor of readability and comprehension.

The methods and solutions presented cover a wide range of engineering problems, illustrated with inspiring examples. Some of them emanate from classical mathematics such as integration, equation solving, and differential equations, and others are cutting-edge topics such as optimization, data mining, genetic algorithms, neural networks, and machine learning. On the other hand, it is not a book for computer experts and a general programming knowledge is sufficient. Even so, an elementary aid to the Python language is included in an appendix to the book.

Since its irruption in the playground of programming languages, Python has earned an honorary place, at the level of Java or C, and its use is intensive in companies and universities. Every day new libraries arise that extend the use of this cool language to almost any scientific or technical discipline. Take as an example the use of Python based libraries such as *Astropy* and *Matplotlib* among others for the composition of the breathtaking photo of the super-massive

1

"Pōwehi" or *The adorned fathomless dark creation* taken from an ancient Hawaiian chant. Black hole at M87 in Virgo Constellation

black hole in the center of the galaxy M87, from many thousands of partial images. As far as I'm concerned, I use Python because there is a very active community of users on the Internet, and above all, ... because it's fun.

I have allowed myself to include images related to the content of the book in one way or another, and I encourage the reader to reflect on their relationship to the themes, sometimes clear but sometimes arguable, sometimes subtle or distant, or simply a personal tribute to people who deserve our admiration.

I have had a great time writing this book and I hope that you also enjoy reading it and that it will be of benefit to you in your studies and professional life.

Javier Riverola Gurruchaga
Madrid, August 2019

Chapter 1

Solving Equations

1.1 Systems of Linear Equations

Many coupled physical systems ideally obey to linear laws. Because of this, linear systems are well known to all students from college courses, and they are present in a multitude of engineering fields. In matrix form, a linear system is:

$$\begin{bmatrix} a_{00} & a_{01} & \dots & a_{0n} \\ a_{10} & a_{11} & \dots & a_{1n} \\ \vdots & & & \\ a_{n0} & a_{n1} & \dots & a_{nn} \end{bmatrix} \begin{Bmatrix} x_0 \\ \vdots \\ x_n \end{Bmatrix} = \begin{Bmatrix} y_0 \\ y_1 \\ \vdots \\ y_n \end{Bmatrix}$$

where $[a_{00}, ..., a_{nn}]$ are linear coefficients, $\{x_0, ..., x_n\}$ are the unknown values, and $\{y_0, ..., y_n\}$ are the given outputs. There is a number of solution methods such as row reduction, elimination of variables, matrix solution, Gaussian elimination, and others.

The following example resolves a linear system using the *numpy linalg* Python library and as an alternative, using the inverse matrix. Obviously, the results are identical.

$$x + 2y + 5z + 6t = 1$$
$$4x - 4y - 6z + 8t = 6$$
$$-12x + y + 3z + 9t = 7$$
$$18x + 6t = 2$$

```python
# System of linear equations

from numpy import array, matmul, linalg

my_matrix = array([ [1,    2, 5, 6],
                    [4,   -4,-6, 8],
                    [-12,  1, 3, 9],
                    [18,   0, 0, 6]])

my_vector = array( [1, 6, 7 , 2])

solution = linalg.solve(my_matrix, my_vector)
print('Solution is x1, x2, x3, x4 =', solution)

# Alternative method

solution1 = matmul(linalg.inv(my_matrix), my_vector)
print('Solution is x1,x2,x3,x4 by inv[A]*y =',
    solution1)
```

```
Solution is x1, x2, x3, x4 =
[-0.18245614  3.27368421 -2.12982456  0.88070175]

Solution is x1,x2,x3,x4 by inv[A]*y =
[-0.18245614  3.27368421 -2.12982456  0.88070175]
```

1.2 Systems of Non Linear Equations

Sooner or later, engineers and scientists face the challenge of solving one or more nonlinear equations whose solution is not immediate because the unknown variables cannot be isolated and the solution cannot be obtained easily by substitution, or another direct method. In these cases we can follow an iterative method in which we start with a first guess and then we approach the solution by successive approximations. The following examples will show the generic use of these iterative routines for systems of one, two, or more nonlinear equations.

1.2.1 Let's go! *Poyekhali!*

Yuri Gagarin aboard Vostok-1 entered space and history for the first time on April 12, 1961. Despite many risks and serious complications, the flight was an unquestionable achievement. His courage, his emotional words during the flight, and his position in favor of world peace in troubled times honor him forever.

I see the Earth! It is so beautiful ... Mankind, let us preserve and increase this beauty, and not destroy it! (Yuri Gagarin)

As a tribute to him, we calculate in this exercise the orbit of Vostok-1 with only two data: the maximum and minimum height (apogee and perigee) reached during his flight. The iterative resolution of the Kepler implicit equation for the implicit *eccentric anomaly* $M(E) = E - e\sin(E)$ justifies its inclusion in this section.

```python
# Vostok1 Orbital flight

from numpy import *
from scipy.optimize import fsolve
import matplotlib.pyplot as plt

# Vostok1 Orbital Data

apogee = 327e3  # m
perigee = 169e3 # m

# Earth data

R = 6378000    # Earth radius, m
M = 5.972e24   # Earth mass, kg
G = 6.672e-11  # Universal Gravity Constant, N m2/kg2

# Ellipse characteristics

mu = M*G
c = (apogee - perigee)/2
a = a = R + apogee -c # major semi axis
b = sqrt(a**2 - c**2) # minor semiaxis
e = sqrt(1-(b/a)**2)  # eccentricity

print("a, b, e ", a, b, e)

# Kepler equation for Mean anomaly

def Ecc_anomaly(E):
    err = t - (a**1.5)*(E-e*sin(E))/mu**0.5
    return err

# Calculation of height vs time

tt = []
hh = []
for t in range(0,60*108,20):
    #
    E = fsolve(Ecc_anomaly,(0))
    r = a*(1 - e*cos(E))
```

```
    tanthetamed = ((1+e)/(1-e))**0.5*tan(E/2)
    theta_rad = 2*arctan(tanthetamed)
    #print("h, theta ",r-R,theta_rad*180/pi)
    tt.append(t)
    hh.append(r - R)
#
ttarray= asarray(tt)
hharray = asarray(hh)

# Now, plot and print results

plt.close('all')
plt.figure(1)
plt.plot(ttarray/60,hharray/1000,0,0)
plt.title('Vostok 1')
plt.xlabel('Time, min')
plt.ylabel('Altitude, Km')
plt.grid()
plt.show()

vmax = sqrt(mu*((2/(R+perigee))-1/a))
print("vmax =",vmax," m/s")

T_minutes = 2*pi*sqrt(a**3/mu)/60
print("T =",T_minutes, "min")
```

```
vmax = 7847.66441335  m/s
T = 89.4783252633 min
```

The script above can be modified to calculate distance traveled, positions on the Earth's sphere, and other orbital parameters.

7

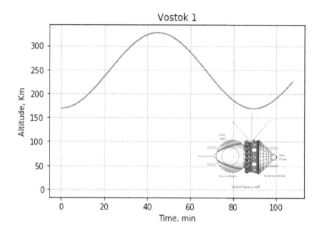

Vostok-1 flight was 108 minutes long, reaching a maximum velocity of 7847 m/s at perigee.

1.2.2 Dear Supernova

It is possible to estimate the time since the Big Bang by taking the inverse of the Hubble constant, which results in the colossal figure of 13.8 billion years. The next cosmic event of great interest is the moment of synthesis of the elements beyond iron within massive stars, and the formation of heavy elements during the explosion of several supernovae and neutron star collisions that gave rise to the residues and cosmic dust from which the planets of our solar system were formed.

Can we estimate how long ago these explosions occurred? Yes, indeed. When heavy elements were formed by nuclear fusion reactions in neutron-rich environments, the slow process of disintegration began, each isotope at its own rate of disintegration.

All uranium isotopes on our planet were created in the course of almost simultaneous stellar explosions on a cosmic scale. Considering that the current proportion of both isotopes is 0.73%, that according to Nuclear Physics the yield of the formation of U-235 is 65% higher than that of U-238, a more massive nucleus, and knowing the constants of disintegration of both isotopes, it is possible to reach the following expression:

$$\frac{N_{235}}{N_{238}} = \frac{N_{235}(0)\exp(-\lambda_{235}T)}{N_{238}(0)\exp(-\lambda_{238}T)}$$

We are star dust (Carl Sagan)

Although the unknown time T can be cleared from this equation, we are going to make an implicit treatment to illustrate the method of nonlinear equations.

```python
# Non linear equations - Supernova Age

from scipy.optimize import fsolve
from numpy import *

# Data

half_life_u235 = 7.038e8 # years
half_life_u238 = 4.468e9 # years

tau_235 = log(2) / half_life_u235
tau_238 = log(2) / half_life_u238

todays_ratio = 0.0073
```

9

```
# Solving the equation

def equation(p):
    T = p
    a = exp(-tau_235*T)
    b = exp(-tau_238*T)
    err1 = todays_ratio - 1.65*a/b
    return err1

T = fsolve(equation, (1000))

print('T =', T/1e6, 'million years')
```

```
T = [ 6533.05238345] million years
```

This value is approximately the accepted one for primordial supernovae explosion. Since then, there have been many more interesting events: the formation of planets, the formation of the atmosphere and the oceans, the irruption of life, ..., and finally the appearance of man. But that is another story.

1.2.3   A message in a bottle!

The previous case is a system with a single implicit equation. In this section we are going to solve a system of two equations, taking as an example a nice problem of terrestrial location.

I was enjoying a nice walk along the shore of the beach on a sunny day, when something bright insistently caught my attention. After wetting my shoes and pants, how wonderful! I recovered a bottle with a touching message inside:

10

After leaving San Francisco, my city of
residence, my flight splashed down in the
Ocean, and after many misfortunes, I reached
an uninhabited island whose location I do not
know! I only have a good watch with which I
have determined that on Easter day, the sun
rised at 8:21 and set at 21:00, Pacific Time. I
hope that with these data you can locate me. If
you are a good-hearted person, please come to
rescue me as soon as possible, before I die of
boredom.
                              Robb Outcast, April 2001

This is a problem with two unknowns: latitude and longitude.
Although it is hard to believe, the castaway has given us enough
data, as we will see hereunder.

We know that the hourly position of the Sun can be well described by the *sun equation*, which can be particularized to sunrise
and sunset as follows:

$$\cos(\phi_r) = \sin(lat)\sin(decl) + \cos(lat)\cos(decl)\cos(ha_r),$$

$$\cos(\phi_s) = \sin(lat)\sin(decl) + \cos(lat)\cos(decl)\cos(ha_s).$$

where ϕ is the zenith angle to vertical, lat is the latitude of the
observer, $decl$ is the declination of sun, and ha is the hour angle
from the local meridian.

The accurate calculation of declination involves very specific parameters like eccentricity, Julian day, and other, as a function of the
fractional year (γ). Besides, conversion between hour angle and
hour needs a correction known as *the equation of time* which also
depends on the fractional year. These equations are clearly explained
in many references, please refer to the National Oceanic and Atmospheric Administration of the USA [1].

---

[1] https://www.esrl.noaa.gov/gmd/grad/solcalc/solareqns.PDF

At sunrise and sunset, the zenith angle is 90 degrees, plus 0.833 degrees due to atmospheric refraction, so we already have two equations that we need to solve.

The Python code below includes the NOAA equations together with the embedded explanations.

```python
# Non linear equations - Outcast

from numpy import *
from scipy.optimize import fsolve

# Some data

sunrise_LST = 8 + 22/60        # sun rise time
sunset_LST = 21                # sun set
time_zone = -8                 # Pacific Time zone
hour = 12                      # reference noon
day_of_year = 31 + 28 +31 + 15

# Fractional year, equation of time and declination

gamma = 2*pi/365*(day_of_year - 1 +(hour - 12)/24)
gamma_d = gamma*180/pi

eqtime_m = 229.18*(0.000075+0.001868*cos(gamma)\
                 -0.032077*sin(gamma)\
                 -0.014615*cos(2*gamma)\
                 -0.040849*sin(2*gamma))

decl = 0.006918-0.399912*cos(gamma)\
       + 0.070257*sin(gamma)\
       -0.006758*cos(2*gamma)\
       +0.000907*sin(2*gamma)\
       -0.002697*cos(3*gamma)\
       +0.00148*sin (3*gamma)

decl_d = decl*180/pi
```

```
# Solve sunrise and sunset for latitude and
    longitude

def solar(p):

    lati,longg = p
    ha=arccos(cos(90.833*pi/180)/ \
            (cos(lati*pi/180)*cos(decl))\
            - tan(lati*pi/180)*tan(decl))
    ha_d = ha*180/pi
    err1 = sunrise_LST -( (720-4*(longg + ha_d)\
            -eqtime_m)/60 + time_zone)
    err2 = sunset_LST - ( (720-4*(longg - ha_d)\
            -eqtime_m)/60 + time_zone)

    return err1,err2

lati,longg = fsolve(solar,(10,-90))

print('lat, lon', lati,longg)
```

```
lat, lon 21.8398685852  -160.189899833
```

Now, we just have to search the latitude and longitude in *Google maps* to locate the island of Niihau, a remote island west of the Hawaiian Islands. Quick, let's save Robb!.

1.2.4 Pressure Drop in a Pipe

Now, an example with three equations and three unknowns. Suppose we have a conveniently instrumented pipe with pressure taps at both ends. As a result, we read a pressure difference of 1000 N/m2. The diameter of the pipe is D = 0.190 m, the length of the pipe L = 2 m, the temperature of the water is T=20 ºC (viscosity is 1004e-6

13

Pa.s, density ρ=0.998 kg/m^3). Can we calculate the flow of water flowing into the pipe?

The problem can be approached in several ways. Let us choose the following set of equations:

$$\Delta P = f_D \frac{L}{D} \rho \left(\frac{\dot{m}}{A\rho} \right)^2$$

$$\frac{1}{\sqrt{f_D/4}} = 4 log_{10} \left[\frac{Re\sqrt{f_D/4}}{1.255} \right]$$

$$Re = \frac{\dot{m}D}{A\mu}$$

```python
# Non linear equations - Pipe

from scipy.optimize import fsolve
from numpy import sqrt, log10

# Data

deltap = 1000
D     = 0.10
L     = 2
mu    = 1004e-6
rho   = 998
A     = 0.00785

# Solving the equations

def equations(p):
    Mdot, fd, Reynolds = p
    err1 = deltap - (1/2)*fd*(L/D)\
                *rho*(Mdot/(A*rho))**2
    err2 = 1/sqrt(fd/4) - 4*log10(Reynolds\
                *sqrt(fd/4)/1.255)
    err3 = Reynolds - Mdot*D/(A*mu)
    return err1, err2, err3
```

```
x, y, z = fsolve(equations, (10, 0.01, 100000))


# Print the solution

print('Solution is:')
print('Mdot =', x, 'kg/s')
print('fd = ', y)
print('Re =', z)
```

```
Solution is:
Mdot = 20.3260787498 kg/s
fd =  0.0148854732663
Re = 257899.342119
```

This type of systems is very common in scenarios of modeling different coexisting phenomena with properties dependent on an external parameter, for example temperature or pressure. Although nature behaves linearly on a very small scale, on the other hand on a large scale different phenomena can interact, which are reflected in gradual or abrupt changes of trend.

1.2.5 One step further!

In this last example we deal with a broader system of nonlinear equations. Higher order systems are resolved identically. It is important to remember that non-linear systems can have several solutions, so it is convenient to try different starting points or first guess values. The system that we propose is the following one:

$$\frac{x^2 + 2}{\sqrt{y}} = 1.1$$

$$xy/2 = 15$$

$$y + zt^2 = 4$$

$$x + y = -z/t$$

$$(xy - zt)u = 0$$

$$v - 30 = 0$$

```
#    System of nonlinear equations

from scipy.optimize import fsolve
from numpy import sqrt

def Equations(p):
    x,y,z,t,u,v = p

    # Equations writen as F(x)=0

    err1 = (x**2 + 2)/sqrt(y) - 1.1
    err2 = x*y/2 - 15
    err3 = y - 4 + z*t**2
    err4 = x + y + z/t
    err5 = (x*y - z*t)*u
    err6 = v - 30
    return err1, err2, err3, err4, err5, err6

x,y,z,t,u,v = fsolve(Equations,[1,1,1,1,1,1])

print('Solution is x,y,z,t,u,v =',x,y,z,t,u,v)
```

```
Solution is x,y,z,t,u,v =
1.64320097451 18.257048567 -17.8066500834
0.894795316054 2.33469435623e-10 30.0
```

1.3   Optimization

Optimization problems arise in a multitude of scientific and technical fields such as economics, physics, biology, and even sociology. In this type of problems we maximize or minimize an objective function f depending on control variables x_j whose domain is either unrestricted or restricted by constraints in the form of inequalities, equations, or both.

The Python based method given here is general and valid for linear and nonlinear functions.

Example of optimization

Minimize $f = x(xy - sqrt(z))$ with the following constraints:

$$x + 2y - 6z < 0$$
$$x - y < 8$$
$$xyz \geq 100$$
$$-10 \leq x \leq 12$$

```
# Optimize with restrictions and bounds

from numpy import *
from scipy import *
from scipy.optimize import minimize
```

17

```python
# Function to minimize

def myfunction(p):
    x,y,z = p
    return x*(x*y-sqrt(z))

# Constraints, rewrite as >=0

def c1(p):
    x,y,z = p
    return -(x+2*y-6*z)

def c2(p):
    x,y,z = p
    return -(x-y)+8

def c3(p):
    x,y,z = p
    return (x*y*z)-100

constraints = [ {'type': 'ineq', 'fun':c1},
                {'type': 'ineq', 'fun':c2},
                {'type': 'ineq', 'fun':c3}]

bounds = ((-10, 12), (-inf, +inf), (-inf, +inf))

# Minimize

results = minimize(myfunction, (1, 1, 1), bounds =
    bounds, constraints = constraints)

print(results)
```

```
    fun: -41.798989870425274
    jac: array([ -3.07106781,  67.94112539,  -0.58284235,
     0.         ])
message: 'Optimization terminated successfully.'
   nfev: 267
    nit: 48
   njev: 46
```

```
   status: 0
  success: True
        x: array([   8.24264069,    0.24264069,   50.          ])
```

The chosen starting point does not always lead to success. Therefore, it is essential to verify that the message "Optimization terminated successfully" has been obtained, and it is frequent that one has to explore the state space to find an adequate starting point.

Proposed exercises

- Solve the following linear system:

$$x + 2y - z = -3$$
$$-2x + y + 4z = 1$$
$$x - y + 2z = 5$$

Ans. $x = 2, y = -1, z = 1$

- Solve the following linear system:

$$x + 2y + z - t = 4$$
$$y + z = 1$$
$$x - 2z - 2t = 3$$
$$z + t = -1$$

Ans. $x = 1, y = 0, z = 1, t = -2$

- Solve the following linear system:

$$\begin{bmatrix} 3 & 5 & 2 \\ 0 & 8 & 2 \\ 6 & 2 & 8 \end{bmatrix} \begin{Bmatrix} x \\ y \\ z \end{Bmatrix} = \begin{Bmatrix} 8 \\ -7 \\ 26 \end{Bmatrix}$$

Ans. $x = 4, y = -1, z = 1/2$

19

- Solve the following non-linear system:

$$x^2 + y = 5$$
$$xy - z = 2$$
$$x\sqrt{y}/z = 1$$

Ans. $x = 1, y = 4, z = 2$

- Solve the following non-linear system:

$$x + yz/t = 2$$
$$t\sqrt{xy} = 2.82$$
$$x + y - z - t = 1$$
$$yz = 0$$

Ans. $x = 2, y = 1, z = 0, t = 2$

- Solve the following non-linear system:

$$xyz = 2$$
$$x + y + z = 4$$
$$xy + yz = 3$$

Ans. $x = 1, y = 1, z = 2$

- Minimize the value of f within the given constraints.

$$f = x + y\sqrt{z}$$
$$x + y \geq 21$$
$$x - z \geq 31$$
$$y + z \geq 52$$

20

Ans. $x = 12, y = 71, z = -19$

- Minimize the value of f within the given constraints.

$$f = x + y - xz$$
$$x + y \geq 18$$
$$x - 2z - 10 \geq 0$$
$$-y - z < 0$$

Ans. $f_{min} = 6, x = 12, y = 6, z = 1$

- Maximize the value of f within the given constraints.

$$f = 10x_1 + 8x_2 + 5x_3$$
$$3x_1 + x_2 \leq 450$$
$$2x_2 + 3x_3 \leq 900$$
$$2x_1 + x_2 \leq 350$$

Ans. $x_1 = 12, x2 = 326, x3 = 82.66, f_{max} = 3141.33$

Let me see: four times five is twelve, and four times six is thirteen, and four times seven is-oh dear! I shall never get to twenty at that rate! Poor Alice, she is doing calculations in base 10 but the answers are coming out in different bases. She is expressing 4n in base 3+3n.

Drawing by Petter Newell, public domain {{PD-US}}

Chapter 2

Engaging with Numbers and Functions

2.1 Interpolation

The general problem of interpolation is to estimate the unknown y value for a given x among pairs of values (x_0, y_0),..., (x_n, y_n). For instance, this might be the case that we require a thermophysical property of a material (conductivity, density, ...) but we have only tables with a few pairs of values, or perhaps it might be the case that we need to estimate a credible result between actual results of a test or calculations.

Interpolation is a kind of routine task but you ought to be careful to avoid unpleasant surprises. There is a suitable method of interpolation for each situation, depending on the risk that you want to assume. This is critical depending on whether you can infer any smooth or abrupt behavior beforehand, or if you know the trends at the extremes of the interpolation interval. We will explore some of these possibilities in this chapter.

2.1.1 Linear Interpolation

According to a conservative strategy, linear interpolation is useful when you do not want to take risks, even at the cost of giving up on obtaining a more accurate prediction. In general, this is the most commonly used type of interpolation.

The idea is quite simple: just find the interval i that contains the x value to interpolate, an draw a straight line passing through both extremes of the interval, so you can estimate the y value as follows:

$$y \simeq f_i + (x - x_i)\frac{f_{i+1} - f_i}{x_{i+1} - x_i}$$

This is easy with Python:
```
y = interp(x, xdata, ydata)
```

We obtain the same result with:
```
y = float(interp1d(xdata, ydata,'linear')(x))
```

The latter form is more sophisticated but more interesting because one can change easily from 'linear' to 'quadratic', or 'cubic', as needed. These options are included in the example below.

2.1.2 Pure Polynomial Interpolation

We search for a polynomial that passes through all n+1 points,

$$P_n(x) = a_0 + a_1 x + a_2 x^2 + \dots + a_n x^n,$$

so we can interpolate y$\simeq P_n(x)$.

As the pursued polynomial must pass through all points, we can write the linear system of equations as a matrix system as follows:

$$\begin{bmatrix} 1 & x_0 & x_0^2 & \cdots & x_0^n \\ 1 & x_1 & \cdots & & x_1^n \\ \vdots & & & & \\ 1 & \cdots & & & x_n^n \end{bmatrix} \left\{ \begin{array}{c} a_0 \\ \vdots \\ a_n \end{array} \right\} = \left\{ \begin{array}{c} y_0 \\ y_1 \\ \vdots \\ y_n \end{array} \right\}$$

and the unknown coefficients can be easily calculated:

$$\{a\} = [X]^{-1} \cdot \{Y\}$$

However, this system is ill conditioned with both small and large numbers together and rounding errors may emerge and ruin the outcome. There are several popular techniques to calculate $\{a_i\}$ without the need to solve the system such as the Newton's, Langrange's, and Aitken's interpolation methods. With Python, this task is immediate:

```
y = poly1d(polyfit(xdata,ydata,len(xdata)-1))(x)
```

It seems tricky but it works. Besides, if we need the values of the polynomial coefficients, then just run:

`a = polyfit(xdata, ydata, len(xdata)-1)` , where $\{a_i\}$ array is ordered from highest to lowest degree term $(a_n, ...a_0)$.

Warnings!

☞ Pure polynomial may oscillate especially near extremes, and near to abrupt changes in data trends. For this reason, a polynomial larger than fifth degree is rarely used. If there are many data points, it is better to perform interpolations based on least squares fit or splines.

☞ In general, extrapolating information beyond the range of data is not recommended. But if necessary, a linear extrapolation is safer than a pure polynomial extrapolation.

2.1.3 Cubic Splines

A spline is a differentiable curve defined in portions by polynomials. It is an interpolation method consisting of dividing the set of points into pieces between which continuity is ensured. This strategy has many advantages and it is consistent with many physical phenomena. We can imagine a spline as a flexible and resistant rod that passes through each of the data points in a smooth manner.

Here, we are going to sketch only the most relevant spline formulation, the *cubic spline with continuous second derivative* (C^2). If our data points are $(x_0, y_0) \ldots (x_n, y_n)$, the spline is a cubic polynomial such as:

$$s(x) = A\frac{(x - x_j)^3}{h_j^3} + B + \frac{(x - x_j)^2}{h_j^2} + C\frac{(x - x_j)}{h_j} + D$$

where x_j is the lower end of the interval containing x, and $h_j = x_{j+1} - x_j$. Parameters A, B, C, and D are obtained by imposing conditions to the first and second derivatives to be the same at both sides of each data point. These conditions are applied to all points except the first and the last one, where boundary conditions are arbitrarily defined, so a system of equations is obtained. Fortunately, Python includes the `scipy.interpolate` module that totally automatizes the process:

```
y = float(interp1d(xdata, ydata,'cubic')(x))
```

2.1.4   Example of Interpolation Methods

This example shows the different methods afore mentioned.

```
#       Interpolation Methods - Example

from numpy import array, poly1d, polyfit
from scipy.interpolate import interp1d

#    Data

xdata = array([2, 3, 4, 6, 12, 18, 22, 33, 40, 45,
    50, 57])
ydata = array([4.5, 10, 16, 37, 120, 100, 83.9, 65,
    64, 66, 70, 71])

xv = 3.5  # value to interpolate

# Linear
yv_lin = interp1d(xdata, ydata,
                  kind='linear')(xv)

# Pure polynomial
yv_pol = poly1d(polyfit(xdata, ydata,
                        len(xdata)-1))(xv)

# Quadratic spline
yv_qspl = interp1d(xdata, ydata,
                   kind='quadratic')(xv)

# Cubic spline
yv_cspl = interp1d(xdata, ydata,
                   kind='cubic')(xv)

print("y_lin =",yv_lin,
"\ny_pol =",yv_pol,
"\ny_qspl =",yv_qspl,
"\ny_cspl =",yv_cspl)
```

```
y_lin = 13.0
y_pol = 12.7535196168
y_qspl = 13.188152688568351
y_cspl = 12.74574808523611
```

Note that the estimates can be very different, depending on whether we are in a tense zone or in a convex zone.

2.2 Approximating Functions

2.2.1 Taylor Series

Since their formal appearance in the early 18th century, Taylor series have been used extensively to study the local behaviors of a function, to calculate limits, to estimate integrals, and to calculate the sum of series, apart from the obvious use of obtaining an approximate polynomial function from a more complicated one.

The formal expression of a Taylor series of a function $f(x)$ in the vicinity of $x = a$, is:

$$f(x) \simeq f(a) + \frac{f'(a)}{1!}(x-a) + \frac{f''(a)}{2!}(x-a)^2 + ... + \frac{f^{(n)}(a)}{n!}(x-a)^n$$

The greater the degree of the polynomial, and the closer we are to the point $x = a$, the more accurate the prediction is. In the attached script we get the series for $f = \sin(x)$. The method is simple and applicable to other functions with no singular points.

```
# Taylor series expansion

from sympy import *
import numpy as np
```

28

```
from sympy.functions import *
import math

# Define the variable and the function to
    approximate

x = Symbol('x')
f = sin(x)

# Set the reference

x0 = 0

# Set the degree of Taylor series

j=5

# Perform the calculations

def taylor(function,x0,n):
    i = 0 ;  p = 0
    while i <= n:
        p = p + (function.diff(x,i).subs(x,x0))\
                /(math.factorial(i))*(x-x0)**i
        i += 1
    return p

func = taylor(f,x0,j)

print('Taylor expansion of f=',f,'\n at n='+str(j),'
    \n',func)
```

```
Taylor expansion of f= sin(x)
at n=5
x**5/120 - x**3/6 + x
```

2.2.2 Chebyshev Polynomials

Unlike Taylor polynomials, which are accurate only in the neighborhood of the central point of the series, approximations of functions based on Chevishev $T_n(x)$ orthogonal polynomials lead to polynomials that are very close to optimal within a closed interval. This is achieved by accumulating interpolating nodes at both ends of the interval. Their high precision makes them very interesting for practical applications

In the attached script we calculate the polynomial of Chebyshev for $f = \sin(x)$ in the interval $[0, \pi/2]$. This powerful method is applicable to many other functions.

```
# Chebyshev Polynomials

from sympy import *
import numpy as np
from sympy.functions import *
from math import factorial

# Set the symbol an function to approximate

x = Symbol('x')
f = sin(x)

# order and interval of Chevishev approximation

n = 4

# Limits of the interval

a = 0
b = pi/2

# Calculate Chebishev Polynomial

sum = 0
```

```
for i in Range(n):
    xi=(a+b)/2 + ((b-a)/2)*cos(pi*(2*i+1)/(2*n+1))
    sum = sum+ f.subs(x,N(xi)) *legendre(i,x)

print(sum)
```

```
0.456790034326054*x**3 + 0.906603197889339*x**2
    + 0.649805511915654*x + 0.696677407683594
```

2.3 Some Interesting Numbers

2.3.1 The Ramanujan Number

The Hardy-Ramanujan number has its origin in a story of friendship between two great men. The famous British mathematician G. H. Hardy went to visit the Indian genius S. Ramanujan, who was hospitalized. Already in his room, he commented that he had written down the taxi number, 1729, in his opinion a "boring number", and added that he hoped it was not a bad omen. "No, Hardy," Ramanujan said, "it's a very interesting number. It is the smallest number expressible as the sum of two positive cubes in two different ways".

An equation has no meaning for me unless it expresses a thought of God (Srinivasa Ramanujan)

From a generalization of this property arise the so-called taxi

31

numbers. In this exercise we will calculate this number from the definition itself.

```
#    Calculation of Ramanujan-Hardy number

click=0
num=1

while (click < 1):
        found = 0
        A = int(num**(1/3))+1
        for alpha in range(1, A):
            B = int(num**(1/3))+1
            for beta in range(alpha + 1, B):
                if (alpha**3 + beta**3 == num):
                    found = found + 1
        if (found == 2):
            click=click+1
            print("Ramanujan number ", num)
        num=num+1
```

```
Ramanujan number    1729
```

2.3.2   The Golden Ratio

The golden ratio is an irrational number with interesting mathematical properties. It is the subject of many brainy disquisitions in the fields of arithmetic, biology, astronomy, and has profound philosophical implications. As a geometric proportion it is present in things as different as the spirals of galaxies, the shell of a snail, the distribution of the leaves of a stem, the rings of a sectioned trunk, or even the crystalline growth. In addition, it is recognizable in the proportions of the canons of aesthetic perfection. The attached script calculates this value, starting from its definition as the limit of the ratios of two consecutive terms of the Fibonacci series.

```python
# Golden ratio

f=[0,1]
for i in range(2,20):
    a = f[i-1]+f[i-2]
    f.append(a)
    fibo = f[i]/f[i-1]

print(fibo)
```

```
1.618034055727554
```

The golden ratio is embedded in multiple chambers, spirals and envelopes of the mysterious Nautilus.

Proposed exercises

- Obtain an estimate of y for x = 6, with hypotheses of linearity, pure polynomial regression, quadratic splines, and cubic splines. The known data are:

x	2	5	7	8	12
y	5	9.2	4.8	9	9.5

Ans. $y_{lin} = 7$, $y_{pol} = 5.1$, $y_{quad} = 4.79$, $y_{cspl} = 5.46$

- Calculate the pure polynomial that passes through the x,y dataset above.

Ans. $y = -0.09x^4 + 2.45x^3 - 22.7x^2 + 83.1x - 88.5$

- Obtain a Taylor polynomial for $f = tan(x)/(1 - x)$, with n=4, and x=0.

Ans. $f \simeq 4x^3/3 + 4x^3/3 + x^2 + x$

- Obtain a Taylor polynomial of $f = cos(x)/sin(x)$, with n=5, and x = 1.5.

Ans.
$f \simeq -1.00502x - 0.13908(x - 1.5)^5 + 0.0478727(x - 1.5)^4 - 0.340063(x-1.5)^3 + 0.0712714(x-1.5)^2 + 1.57845$

- Obtain a third degree Chevishev polynomial of $f = tan(x)/(1-x)$, inside $[-\pi/2, \pi/2]$

Ans. $f \simeq -2.94728160x^3 - 0.329681439x^2 +$
$6.42816132x - 21.9976721$

- Obtain a fourth degree Chevishev polynomial of
  $f = \cos(x)/\sin(x)$, inside [0,3]

Ans. $f \simeq 18.0245346x^4 + 2.07986450x^3 -$
$15.6650790x^2 - 2.5382639x - 3.25743228$

*We are just an advanced breed of monkeys on a minor planet of
a very average star. But we can understand the Universe. That
makes us something very special.* (Stephen Hawking)

Chapter 3

Integration

Integral calculus is present in virtually any physics or math book. Finding the primitive form of a function often has been a non-negligible challenge. Since the first college courses, the appearance of these pitfalls in exercises or exams, has put us in trouble. But in reality, the integral is one of the most powerful concepts of Calculus, and the search for resolution methods has captivated more than one clear mind such as Newton himself, Leibnitz, and Riemann among others. In this chapter we

If I have seen further it is by standing on the shoulders of Giants. (Isaac Newton)

will learn that the solution of integrals of all kinds with Python is easy and indefinite, improper, and even multidimensional integrals will no longer be a problem for us.

3.1 Indefinite Integral

The calculation of primitives of rational and trigonometric expressions is tricky, and usually requires a change of variable. The result is another expression of the same nature. Python includes the SymPy module and the integrate function, which in a simple way automates the calculation of the primitive.

Example. Find the primitive of the following indefinite integral

$$\int \frac{1}{u^2 - a^2} du$$

```python
# symbolic integration of indefinite integral

from sympy import *

u = Symbol ("u")
a = Symbol("a")

ii = integrate (1/(u**2-a**2), u)

print ("Result is ", ii)
```

```
Result is  (log(-a + u)/2 - log(a + u)/2)/a
```

Note that u and a have been declared as symbolic variables, since this is a necessary step before performing any symbolic operation. In the lines that follow, the reader is invited to solve more symbolic integrals whose resolution by hand is somewhat tedious but with Python is immediate. The mechanics is exactly the same as in the example above.

Proposed exercises

- Solve $\int \cos(u)\sin(u)du$

  Ans. $\sin(u)^2/2 + C$

- Solve $\int e^u(u^2 - 1)du$

  Ans. $e^u(u^2 - 2u + 1) + C$

- Solve $\int (u^2 - \frac{6u}{u^2+2})du$

  Ans. $u^3/3 - 3\log(u^2 + 2) + C$

- Solve $\int \frac{1}{\sqrt{u^2+a^2}}du$

  Ans. $arcsinh(\frac{u}{a}) + C$

Note that C integration constant has been added in all cases.

3.2 Improper Integral

This class of integrals has one or two limits that approximate a real number or infinity. There are several techniques to solve improper integrals, such as calculation of residuals or other. Next example ilustrates a Python based method.

Example. Solve the following improper integral

$$\int_0^\infty \frac{dx}{1 + x^2}$$

This integral can be solved by first integrating in definite limits [0,b] and then calculating the limit b → ∞, or it can also be considered as a Lebesgue integral over the interval $[0, \infty]$. In this example, we take advantage of the simplicity of Python in just a few lines as follows.

```python
#    Improper integral

from scipy.integrate import quad
from scipy import *

Fun = lambda x : 1/(1+x*x)
I = quad (Fun , 0, inf)
print ("Result is ", I )
```

```
Result is  (1.5707963267948966, 2.5777915205989877e-10)
```

The result of this operation is an array with two values, being the first the value of the improper integral ($\pi/2$), and the second an estimate of the error.

Proposed exercises

- Solve $\int_0^\infty \frac{dx}{\sqrt{x}}$

Ans. 2

- Solve $\int_0^\infty e^{-x^2} dx$

Ans. $\sqrt{\pi}/2$

3.3 Areas

3.3.1 Area between a curve and $x-$axis

Occasionally, we need to find the area under a curve. Normally, to integrate this curve between the limits is enough, but we must be careful to consider only the positive part of the area to avoid the mutual cancellation of the areas above and below the axis.

Example. Find the area between $y = x^3 - 7x^2 + 10$ and x-axis from x=0 to x=5.

$$\int_0^5 \mid x^3 - 7x^2 + 10x \mid dx$$

This curve is positive from 0 to 2, and turns negative from 2 to 5. Therefore, it is necessary to bracket it as an absolute function; otherwise both areas would be mutually canceled.

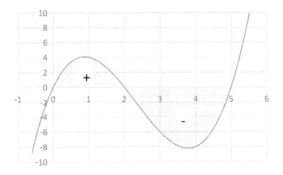

```
#    Area of a function

from scipy.integrate import quad
from scipy import *
from numpy import *

Fun = lambda x : abs(x**3-7*x**2+10*x)
Area = quad (Fun , 0 , 5)
print ("Area curve to x-axis is ",Area )
```

```
Area curve to x-axis is
(21.083333334213776, 2.2229220064673427e-07)
```

3.3.2 Area between two curves

Example. Find the area between $y = 3x - x^2$ and $y = x$.

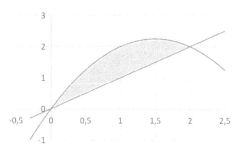

This case is trivial if one realizes that both curves can be subtracted.

$$S = \int_a^b f_2(x) - f_1(x)dx = \int_0^2 \left(3x - x^2\right) - xdx$$

```
# Area between two curves

from scipy.integrate import quad
from scipy import *
from numpy import *

Fun = lambda x : 3*x - x**2 - x
S = quad (Fun, 0 ,2)
print ("Area between two curves is  ",S )
```

```
Area between two curves is
(1.3333333333333333, 1.4802973661668752e-14)
```

3.4 Arc Length

The arc length of a curve between two given abscissas, a and b, can be calculated as follows:

$$L_a = \int_a^b \sqrt{1 + \left(\frac{dy}{dx}\right)^2} \, dx$$

The term with the derivative inside the formula could be a nuisance, but Python has a specific function for it.

Example. Find the arc-length of $y = x^2$ between $x = 0$ to $x = 1$.

```
#   Arc-length

from scipy.misc import derivative
from scipy.integrate import quad

Fun = lambda x: x**2
```

43

```
lon = lambda x: sqrt(1+derivative(Fun,x,1e-5)**2)
arc_length1 = quad (lon, 0, 1)
print ("La is ", arc_length1 )
```

```
La is
(1.478942857542138, 1.402135532654134e-13)
```

Example. Find the arc-length of a cycloid given by $x = 5(\theta - sin(\theta))$, $y = 5(1 - cos(\theta))$, between $(0 \leq \theta \leq \pi)$. Note that this curve is defined as both x and y dependent of parameter θ.

```
#    Arc-length

from scipy.misc import derivative
from scipy.integrate import quad

xFun =    lambda t : 5*(t-sin(t))
yFun =    lambda t : 5*(1-cos(t))

lon =    lambda t : (derivative(xFun,t, 1e-5)**2 +
         derivative(yFun, t, 1e-5)**2)**0.5

arc_length = quad (lon, 0, 2*pi)
print ("La is  ", arc_length )
```

```
La is
(39.99999999917042, 4.440892098408524e-13)
```

Proposed exercises

- Find the arc length of $y = x^3 - 7x^2 + 10x$, from $x = 0$ to $x = 5$.

44

Ans. $L_a = 25.4$

- Find the arc length of the catenary curve $y = cosh(x)$, from $x = 0$ to $x = \ln(2)$.

Ans. $L_a = 3/4$

- Find the arc length of the curve $y = ln[sec(x)]$, from $x = 0$ to $x = \pi/4$.

Ans. $L_a = ln\left(\sqrt{2} + 1\right)$

3.5 Volume of Revolution

$$V_x = \pi \int_a^b [f(x)]^2 \, dx$$

Example. Find the volume of revolution of $y = 2x^2$ between $x = 1$ to $x = 3$.

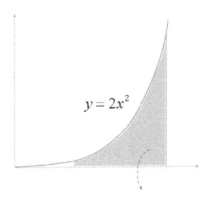

$y = 2x^2$

```
#    volume of revolution

from scipy.integrate import quad
from scipy import *
from numpy import *

Fun    = lambda x : 2*x**2
vol0   = lambda x : pi*Fun(x)*Fun(x)
volume = quad (vol0, 1, 3)

print ("Volume of rev around x-axis is ", volume )
```

```
Volume of rev around x-axis is
(608.2123377349841,  6.752513412144713e-12)
```

If the body is rotated about the y-axis rather than the x-axis, the we use:

$$Vy = \pi \int_a^b x^2 dy$$

where $x = f^{-1}(y)$.

Proposed exercises

- Find the volume of revolution of y=5 around x-axis, from $x = 1$ to $x = 3$.

Ans. $V = \pi 5^2 4$

- Find the volume of revolution of $x^2 + y^2 = 10^2$ around x-axis, from $x = 0$ to $x = 10$.

Ans. $V = 4\pi 10^3/3$

3.6   Moment of inertia

The moment of inertia around the y axis is calculated as follows:

$$I_{yy} = \int_a^b x^2 f(x) dx$$

Example. Find the moment of inertia of $y = 4 - x^2$ around y-axis.

```
#    Moment of Inertia around yy
#
from scipy.integrate import quad
from scipy import *
from numpy import *

Fun  = lambda x : 4-x**2
Fun1 = lambda x : x**2*Fun(x)

Iyy = quad (Fun1, -2, 2)
print ("Iyy is   ",Iyy)
```

```
Iyy is    (8.533333333333331, 9.473903143468e-14)
```

On the other hand, when the moment of inertia is around the x axis, a change of integration base must be made.

$$I_{xx} = \int_a^b y^2 g(y) dy$$

Example. Find the moment of inertia of $y = 4 - x^2$ around x-axis.

```
# Moment of inertia around xx

from scipy.integrate import quad
from scipy import *
from numpy import *

Fun = lambda y  : sqrt(4-y)
Fun2 = lambda y : 2*y**2*Fun(y)

Ixx = quad (Fun2, 0, 4)
print ("Ixx is  ",  Ixx)
```

```
Ixx is
(39.00952380952331, 1.2706752983149272e-07)
```

3.7   Double and Triple Integrals

The double and triple integrals require maintaining the order of integration while defining the functions and limits. The following examples illustrate the method.

Example. Calculate $\int_{y=0}^{1} \left(\int_{x=0}^{2} xy^2 dx \right) dy$

```
#    Double Integral

from scipy.integrate import dblquad

f = lambda x, y: x*y**2
# x,y above same order of integration as dx dy

I_box = dblquad(f, 0, 1, lambda x: 0, lambda x: 2)
# outer limits, then inner limits
```

```
print('Result is ', I_box)
```

```
Result is
(0.6666666666666667, 2.2108134835808843e-14)
```

Example. The limits can be variable expressions:

$$\int_0^1 \int_y^{y^2+1} x^2 y \, dx \, dy$$

```
#   Limits are Variable Expressions - Example

from scipy.integrate import dblquad

I_9 = dblquad(lambda x,y : x*x*y,
              0, 1,
              lambda y: y, lambda y: y*y+1)

# x,y above same order of integration as dx dy
# outer limits, then inner limits

print('Result is I_9 =',I_9)
```

```
Result is I_9 =
(0.5583333333333333, 2.56811395902259e-14)
```

Example. We can also integrate with reverse order dy dx.

$$\int_{x=0}^1 \int_{y=0}^{e^x} \left(x + y^2\right) \, dy \, dx$$

49

```
from scipy.integrate import dblquad

I_7 = dblquad(lambda y,x : x + y**2, 0, 1, lambda x:
    0, lambda x: exp(x))

# y,x above same order of integration as dy dx
# outer limits, then inner limits

print('Result is I_7 =',I_7)
```

```
Result is I_7 =
(3.1206152136875187, 1.038969199265395e-13)
```

Example. A tripple integral :

$$\int_{x=1}^{2} \int_{y=0}^{x^2} \int_{z=0}^{2-x-y} \left(x^2 + y^2 + z^2\right) \, dz \, dy \, dx$$

```
#    Triple Integral

from scipy.integrate import tplquad

f = lambda z, y, x: x**2 + y**2 + z**2
ext1 = 0
ext2 = 2
mid1 = lambda x: 0
mid2 = lambda x: 2-x
int1 = lambda x, y: 0
int2 = lambda x, y: 2-x-y

I_box3 = tplquad(f,
                 ext1, ext2,
                 mid1, mid2,
                 int1, int2)
print(I_box3)
```

```
(1.6, 2.935005303331644e-14)
```

3.8   Integration of Data Sets

In engineering, it is common that we do not have an analytical function, but only a set of (x,y) pairs of values. This is the normal case of tables of mechanical, thermophysical, or other properties. The integral can not be done with the previous methods and it is necessary to use the so-called *numerical quadratures*, in which the integral is replaced by a weighted sum:

$$\int_a^b [\{X\} \to \{Y\}] \cdot dx \simeq \sum_0^n w_i \cdot f(x_i)$$

Some quadratures are based on interpolation and others based on adaptive polynomials. Among the first, the most known and used are those of sum of rectangles by intervals, sum of trapezoids by intervals, Simpson's rule (2nd degree polynomials in equispaced points), Newton-Cotes quadratures (equispaced trapezoids), and Gaussian quadratures (variable integration points). The use of these quadratures is appropriate when the *integration points or nodes* coincide with data, or when data are equispaced, but this is not always the case. Another difficulty that may arise is that the integration limits may not coincide with any of the known abscissas x_i.

The simplest way is to assume an interpolation function that passes through all the points and integrate this function in the interval with one of the previously exposed techniques. In the example of this section, we will assume linear interpolation and also splines interpolation for comparison purposes, and we will see that the result may differ. We must discern always which is the behavior that

51

best fits our data. In general,unless there is a compelling reason, the linear interpolation is safer.

```python
#    Integration of data sets

from scipy.integrate import quad
from scipy.interpolate import interp1d
from numpy import array

# My data points

x1 = array([100, 130, 170, 190, 230, 270, 320, 370])
y1 = array([.598, 1.496, 4.122, 6.397, 13.99, 28.09,
    64.72, 203.0])

# Integrate assuming linearity

I_linear = quad(lambda x: interp1d(x1, y1, 'linear')
    (x), 150, 300)

# Integrate assuming smoothness (splines)

I_spline = quad(lambda x: interp1d(x1, y1, 'cubic')(
    x), 150, 300)

print("Ilinear =", I_linear)
print("Ispline =", I_spline)
```

```
Ilinear = (2596.210000517, 1.86111856237402e-06)
Ispline = (2434.506338741, 2.93566422165013e-05)
```

3.9 Fourier Transform

The integral known as the Fourier Transform deserves special attention for its applications in various branches of science. One of the most widespread uses is to decompose a time signal $f(t)$ into its frequency components. The analysis of time signals and their decomposition into frequencies allows us to reveal hidden characteristics of the signals. It is widely used in sound compression processes, analysis of the response of complex dynamic systems, not to mention the study of signals received in radio telescopes.

$$\widehat{f}(\omega) = \frac{1}{\sqrt{2\pi}} \int_{-\infty}^{\infty} f(t)e^{i\omega t}$$

Often, the time signal f(t) is a discrete sample, so the *Discrete Fourier Transform* is applicable.

$$y_k = \sum_{n=0}^{n-1} \exp(-2\pi j \frac{kn}{N}) f_n$$

The following example is a good exercise of applying the Fourier transform of a signal in the time domain to break it down into its main frequencies.

3.9.1 Listen to that whale!

On the internet you will find many interesting sounds. In this exercise we analyze the song of humpback whales. The reader is invited to download the file "whale1a.wav",

53

available on the website of NOAA (National Oceanic and Atmospheric Administration). Note that, before performing the Fourier analysis itself, we apply a Hann filter or window to dampen the aliasing effect of the ends.

We also suggest the reader to analyze other interesting sounds such as "LaughingChild.wav" (a child's laughter), "rwbl10.wav" (blackbird song), and "thunder3.wav". To download them just enter these IDs in Google.

```python
# Humpback whale calling

from scipy.io import wavfile
import numpy as np
from matplotlib import pyplot as plt
import os

# Read the wav file

filein = 'whale1a.wav'
samplerate, data = wavfile.read(filein)
N = len(data)

# set a times axis, for plotting

times = np.arange(len(data))/float(samplerate)
dt=times[1]-times[0]

# Plot the wave

plt.close('all')
plt.figure(1)
plt.plot(times, data)
plt.title('Time Domain Signal')
plt.xlabel('Time, s')
plt.ylabel('Amplitude ($Unit$)')

# Set and apply a Hann window

hann = np.hanning(len(data))
```

```
data_hann = hann*data   # apply Hann

# Perform FFT

YY = np.fft.fft(data_hann) # Calc FFT
f = np.linspace(0, samplerate, N, endpoint=True)

# Normalize to /N

YYnorm = YY*2/N

# Discard 2nd half

fhalf = f[:int(N/2)]
YYnormhalf = YYnorm[:int(N/2)]

# Plot spectrum

plt.figure(2)
plt.plot(fhalf, abs(YYnormhalf))
plt.title('Frequency Domain Signal')
plt.xlabel('Frequency ($Hz$)')
plt.ylabel('Amplitude ($Unit$)')

# Print some results

print("\nAnalysis of "+filein)
print('number of data N=',N)
print('Sampling dt=',dt)
print('Fmin Hz=',fhalf[0],' Fmax Hz=',fhalf[-1])
```

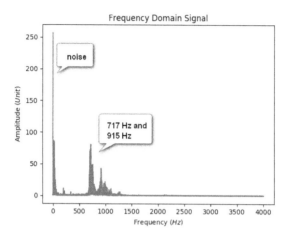

Whale calling in frequency domain. Two peaks emerge at 717 Hz and 915 Hz.

3.10 Montecarlo Integration

Now, some fresh air! Let us calculate number π with a Montecarlo integration.

We know that the area inside a circle of radius r=1 is just the number π. Therefore, we may perform the following integral:

$$\pi = \iint_\Omega dx\, dy$$

being Ω a circle of prescribed radius. To this end, we generate a very large number of trials inside a box of sides x[-1 to 1] and y[-1 to 1], and compute success if $x^2 + y^2 < 1$.

56

```python
# Calculate pi with Montecarlo

from numpy import random

# bounding box
x1, x2= -1, 1
y1, y2= -1, 1

# generate x,y samples
n = 500000 ; sum = 0

x_sample = random.uniform(x1, x2, n)
y_sample = random.uniform(y1, y2, n)

for i in range(0,n):
    x = x_sample[i]
    y = y_sample[i]

    # check if x,y inside bounds
    if (x**2 + y**2 < 1): inside = 1
    else: inside = 0

    # accumulate function when inside
    fun = 1
    sum = sum + fun*inside

I_montec=sum/n*(x2-x1)*(y2-y1)
print('PI_Montecarlo = ', I_montec)
```

```
PI_Montecarlo =   3.141488
```

which is close to the true value of $\pi = 3.14159265359...$ Although the sample has a large number of cases, the result will never be exact. Every time we run the script we get a different estimate of π, always around the real value.

Nothing in life is to be feared, it is only to be understood. Now is the time to understand more, so that we may fear less. (Marie Sklodowska Curie, first person to win the Nobel prize twice)

Chapter 4

Differential Equations

Differential equations are of great importance in science and engineering, because many physical relationships such as laws of conservation, laws of dynamics, of electromagnetism, of physics and chemistry, and many others, appear mathematically in the form of differential equations. There are homogeneous and non-homogeneous, linear and non-linear, ordinary and partial derivatives differential equations.

4.1 Ordinary Differential Equations

It is a differential equation with only one independent variable, usually time. In order to solve a set of ordinary diferential equations, we write it in the form:

$$y^{(n)} = F(t, y, y^{(1)}, ..., y^{(n-1)}).$$

The study of ordinary differential equations (ODE) and their solution is a matter of complete courses at college, a discipline of

Calculus that is approached with different techniques such as direct integration, variable separation, linearization, Fourier series and Laplace transformations, and also with numerical methods such as Euler and Runge-Kutta. In the following examples we are going to use the scipy.integrate Python library.

4.1.1 A First Example

In this first case, we are going to integrate a non-linear ODE as an initial value problem and then draw the phase diagram. Already in this first example, the reader will perceive the power and simplicity of the method exposed here.

$$\frac{dx}{dt} = x - y + \frac{xy}{1000}$$

$$\frac{dy}{dt} = 6x - 2y + 9$$

```python
#    System of Differential Equations

from scipy.integrate import odeint

from numpy import arange

# set a time scale

t = arange(0, 15, 0.1)

#    Define the ODE system

def derivatives(state,t):
    x, y = state
    x_dot = x - y + 2 + x*y/100
    y_dot = 6*x - 2*y + 9
    return [x_dot, y_dot]
```

```
#   Solve the ODE system

solution_x_y = odeint(derivatives, [2, 3], t)

#   Unpack the solution

x = solution_x_y[:, 0]
y = solution_x_y[:, 1]

#   Draw phase plot and run plot

import matplotlib.pyplot as plt
plt.close('all')

fig = plt.figure(1)
plt.plot(x[0], y[0], 'o', x, y, 'r')

plt.title('Phase Diagram')
plt.xlabel('x')
plt.ylabel('y')
plt.legend(('start', 'run'))

fig = plt.figure(2)
plt.plot(t, x, '--', t, y)
plt.title('x and versus time')
plt.xlabel('time, s')
plt.ylabel('x and y')
plt.legend(('x','y'))
```

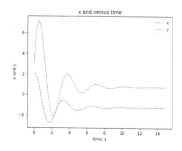
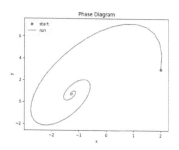

x and y run plots and phase diagram

4.1.2 The Buttered toast mystery!

Murphy's law, quintessence of pessimism, states that "If something bad can happen, it will happen". However, this statement deliberately ignores that there are many daily events with a good outcome; just have your eyes wide open to honestly notice it. One of the classic examples used to illustrate Murphy's law is the *mystery of toast* that tends to land butter-side down when ot falls. Actually, there are physical reasons for this outcome and has been the subject of interesting studies, such as the nice paper *Which side up? Falling bread revisited*, by Koupil and Dvorak[1], wich includes the dynamic equations of motion:

---

[1] http://kdf.mff.cuni.cz/˜janek/pocitace/falling_bread.pdf

$$F_p = \frac{J_c}{Jc + mr^2} m \left(g \cos \phi - 2\dot{r}\dot{\phi} \right)$$

$$\ddot{\phi} = \frac{r}{J_c} F_p$$

$$\ddot{r} = -\frac{F_f}{m} + r\dot{\phi}^2 + g \sin \phi$$

being F_p, the normal force on the toast, F_f, the friction on the toast lower surface, J_c, the moment of inertia, m, the toast mass, r, distance of c.o.g. to the contact corner of the table, and ϕ, the angle of the toast at any moment.

```python
# Falling bread

from scipy.integrate import odeint
from numpy import arange, cos, sin, argmax, pi
import matplotlib.pyplot as plt

# Some data in SI of Units

m = 0.1      # loaf mass
L = 0.11     # length of loaf
r0 = 0.01    # initial c.o.g. displacement
f = 0.25     # friction
H = 0.75     # Height of table
g = 9.81     # gravity

Jc = (1/12)*m*L**2

# set a time scale

t = arange(0, 5, 0.001)

# Define the ODE system

def derivatives(state,t):
    phi, om, r, vr = state
    #
```

```
    Fp = (Jc/(Jc+m*r**2))*m*(g*cos(phi)-2*vr*om)
    if Fp<0: Fp=0
    Ff = f*Fp
    #
    phi_p = om
    r_p = vr
    om_p = r*Fp/Jc
    vr_p = -Ff/m + r*om**2 + g*sin(phi)
    return [phi_p, om_p, r_p,vr_p]

# Solve the ODE system

solutions = odeint(derivatives, [0, 0, r0, 0], t)

# Unpack solutions

phi =    solutions[:, 0]
om =     solutions[:, 1]
r =      solutions[:, 2]
vr =     solutions[:, 3]

Fp = (Jc/(Jc+m*r**2))*m*(g*cos(phi)-2*vr*om)
xc = r*cos(phi)
yc = r*sin(phi)

# Position of the toast when yc in landing range

pos1=argmax(yc<-H+L/2)
phi1 = phi[pos1]

res="unknown"
if phi1 > pi/2:
    if phi1 < 3*pi/2:
        res="buttered down :-("
    else:
        res="buttered up :-)"
print(res)
```

When the center of the toast is in the landing position, if the angle ϕ is between $\pi/2$ and $3\pi/2$, the butter side is down. If we run this script a large number of times with different initial values, in the vast majority of cases the toast falls down the butter side, which is in line with popular belief. Thus, the annoying behavior of the toast is due to the inexorability of the laws of mechanics, obedient in their journey of only 0.75 meters high. But note, if the table were higher, say 1.5 meters high, the toast would always land right!

4.1.3   Complex Pendulum

In this example we assume that we have a ruler hanging from a hole near one end and let it swing like a pendulum. This problem is often presented as a case study of solid dynamics because it allows the validity of mechanical equations to be illustrated in a simple way with a reproducible elementary experiment: it is enough to have a real ruler, a fixed point and a stopwatch. After establishing the balance of forces it is easy to arrive at the following equation of motion:

$$\frac{d^2\theta}{dt^2} = -\frac{mgx}{I_0}\sin(\theta)$$

being $I_0 = mL^2/12 + mx^2$, the moment of inertia of the rule around the fixed point.

The aim is to visualize the free oscillation of this rule and compare the period of oscillation with the theoretical one $T = 2\pi\sqrt{\frac{I_0}{mgx}}$.

```
# Oscillating ruler

from scipy.integrate import odeint
from numpy import *

# Data

L = 0.46    # m
```

65

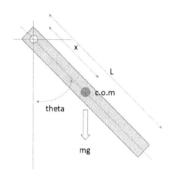

```
m = 0.042    # m
x = 0.18     # m
g = 9.81     # m/s2

I0 = m*(L**2/12) + m*x**2

# Set a time scale

t = arange(0, 5, 0.05)

# Define the ODE system

def deriv(state, t):
    psi, theta = state
    psi_dot = -m*g*x*sin(theta)/I0
    theta_dot = psi
    return [psi_dot, theta_dot]

# Solve the ODE system

soluc = odeint(deriv, [0,pi/4],t)

# Unpack the solution

psi = soluc[:, 0]
theta = soluc[: ,1]
```

```
# Plot results

import matplotlib.pyplot as plt
plt.close('all')
plt.plot(t, theta, t, psi)
plt.legend(('Theta', 'Psi'))
plt.title('Complex Pendulum')
plt.xlabel('time, s')
plt.ylabel('Psi, theta, rad')
```

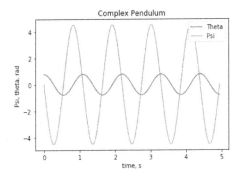

The time lapse between two peaks is 1.1 sec. It is the same time as calculated with the anlytical period

$$T = 2\pi\sqrt{\frac{I_0}{mgx}} = 1.1 \text{ sec}$$

67

4.1.4   The Butterfly Effect

The butterfly effect first appears in the science fiction story *The Thunder* by Ray Bradbury in 1952. It shows that under certain circumstances, a small variation of the initial conditions can have great

consequences in the medium or long term. It is one of the paradigms of chaos theory. In 1963, Edward Lorenz developed an atmospheric model of an oscillating solution known as the Lorenz Attractor. It was found by chance that for certain combinations of σ (Prandtl number), and ρ (Rayleigh number), the oscillations of the attractor lead to chaotic behavior. In addition, the phases or trajectories of the solution describe clear butterfly wings, so it has become the perfect example to illustrate this effect. The Lorentz attractor equations are:

$$\dot{x} = \sigma(y - x)$$
$$\dot{y} = x(\rho - z) - y$$
$$\dot{z} = xy - \beta z$$

In this exercise we will solve these equations from a chaotic configuration ($\sigma = 10$, $\rho = 28$ and $\beta = 8/3$), starting from the point [0.0, 0.0001, 0.0] and with a time step small enough for the wings to appear.

Curiously, if starting from slightly different initial conditions (ie [0.0, 0.0002, 0.0]), the solution obtained is almost identical during the first 30 seconds of simulation, but after that moment, the trajectory follows a remarkable bifurcation. The emergence of this abrupt temporal phenomenon gives rise to interesting philosophical speculations.

```python
# The Butterfly Effect

from numpy import *
from scipy.integrate import quad, odeint
import matplotlib.pyplot as plt

# Data

sigma = 10
rho = 28
beta = 8/3

# Set a time scale

t = arange(0, 200, 0.05)

# Define the ODE system

def derivatives(state,t):
    x,y,z = state
    xdot = sigma*(y-x)
    ydot = x*(rho - z)-y
    zdot = x*y - beta*z
    return [xdot, ydot, zdot]

# Solve the ODE system

sols = odeint(derivatives, [0, 0.0001, 0], t)

# Unpack the solution

x = sols[:, 0] ; y = sols[:, 1] ; z = sols[:, 2]

# Results

from mpl_toolkits.mplot3d import Axes3D
plt.close('all')
fig = plt.figure(1)
ax = plt.axes(projection='3d')
ax.plot(x,y,z)
```

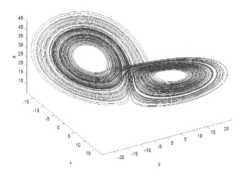

The solution of the Lorentz attractor deploys as butterfly wings. This plot is extremely sensitive to very small departure from the initial condition.

4.1.5 Dancing Planets

After reviewing the meticulous observations of Tycho Brahe, Johannes Kepler surpassed himself by abandoning his belief in the Theory of Harmony of the Celestial Spheres. He realized that the planetary movement does not fit circles or ovals, but only ellipses. In 1609 he published his famous three laws being recognized immediately as the best astronomer of his time.

Kepler crater on the moon (NASA)

In this exercise we will verify the fulfillment of the three laws of Kepler's planetary motion, establishing the equilibrium of forces on the planet Mars according to the Law of Universal Gravitation and the Fundamental Law of Mechanics, both due to the great Isaac Newton. We will calculate the trajectory of the planet taking as initial condition the furthest point from the Sun. The resulting dynamic equations of motion are:

$$\ddot{x} = -\frac{GMx}{r^3}$$

$$\ddot{y} = -\frac{GMy}{r^3}$$

being x, and y, the Sun centered coordinates of the planet, r, is the distance to Sun, G, is the gravity constant, and M the mass of the planet.

```
# Planetary motion

from scipy.integrate import odeint
from numpy import *


# Mars Data

G = 6.67408e-11          # Gravity, km3/s/kg
UA = 1.49600e8*1000      # Astronomic Unit in meters
a = 1.523705*UA          # semi-major axis
e = 0.093404             # eccentricity
M = 1.989e30             # mass, kg
velocity = 21958.3       # m/s
aphelion = 1.66602514182*UA


# Set a time scale

tmax = 2*365*24*3600     # two years in seconds
dt = 500                 # time step, sec
t = arange(0, tmax, dt)  # set a time axis


# Define the ODE system

def derivs(state, t):
    #
    x, y, vx, vy = state
```

71

```
    r = (x**2 + y**2)**0.5
    #
    vxdot = -G*M*x/r**3
    vydot = -G*M*y/r**3
    xdot = vx
    ydot = vy
    return [xdot, ydot, vxdot, vydot]

# Solve the ODE system

init_cond = [-aphelion, 0, 0, velocity]
solutions = odeint(derivs, init_cond, t)

# Unpack the solution

x  = solutions[:,0]
y  = solutions[:,1]
vx = solutions[:,2]
vy = solutions[:,3]


tyear = t/(365*24*2600)

#   Polar coordinates

r = (x**2 + y**2)**0.5
theta = arctan2(y, x)*180/pi

# sweep areas versus time

da_dt = 0.5*diff(theta)/diff(t)*r[0:-1]**2
variation_over_mean = std(da_dt[0: 100000])\
                      /mean(da_dt[0:100000])*100
print('sigma(da_dt) / mean(da_dt) %',
      variation_over_mean)

# Draw phase plot and run plot

import matplotlib.pyplot as plt
plt.close ('all')
fig = plt.figure (1)
plt.plot (x/UA,y/UA,'-')
```

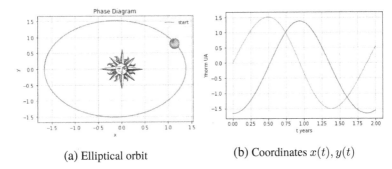

(a) Elliptical orbit

(b) Coordinates $x(t), y(t)$

Plots of Mars trajectory

```
plt.title ('Phase Diagram')
plt.xlabel ('x')
plt.ylabel ('y')
plt.legend (('start','run '))
plt.grid(True)

fig = plt.figure (2)
plt.plot(t/(365*24*3600),x/UA,t/(365*24*3600),y/UA)
plt.xlabel ('t years')
plt.ylabel ('Ynorm UA')
plt.grid(True)
```

Kepler's First law: *The orbit of a planet is an ellipse with the Sun at one of the two foci.*

In Figure (a) we see that the trajectory of the planet describes an ellipse, and the Sun is precisely located at point (0,0).

Second Law: *A line segment joining a planet and the Sun sweeps out equal areas during equal intervals of time.*

This is verified by seeing that the variation to mean ratio (0.00038 %) is an insignificant value, and it is not zero due to the limited precision of the calculation.

Third Law: *The square of the orbital period of a planet is directly proportional to the cube of the semi-major axis of its orbit:* $\frac{a^3}{T^2} \simeq 7.496x10^{-6}\frac{AU^3}{days^2}$ *is a constant value, the Kepler's constant.*

We can verify this point by measuring the period between two peaks in Figure (b), resulting T=1.88 UA, and performing the calculation which results in $7.662x10^{-6}$. Close enough!

Proposed exercises

- The **Lotka-Volterra** equations predict the evolution of an ecological predator-prey dual system in a closed world:

$$\dot{x} = Ax - Bxy$$
$$\dot{y} = -Cy + Dxy$$

  Solve these equations for the following parameters: A = 1.5 (the growth rate of prey), B =1 (the rate at which predators destroy prey), C =3 (the death rate of predators), and D = 1 (the rate at which predators increase by consuming prey), with initial conditions x = 10 (prey), and y = 10 (predator).

- Solve the **Continuous-time logistic growth** equation:

$$\frac{dN}{dt} = rN\left(1 - \frac{N}{K}\right)$$

  where K = 100 (population carrying capacity), r = 3 (growth rate), and $N_0 = 1$.

- Solve the **double pendulum** equations with arbitrary lengths, masses, and initial conditions θ_1 and θ_2:

$$(m_1 + m_2)l_1\ddot{\theta}_1 + m_2l_2\ddot{\theta}_2\cos(\theta_1 - \theta_2)+$$
$$m_2l_2\dot{\theta}_2^2\sin(\theta_1 - \theta_2) + g(m_1 + m_2)\sin\theta_1 = 0$$

$$m_2 l_2 \ddot{\theta}_2 + m_2 l_1 \ddot{\theta}_1 \cos(\theta_1 - \theta_2) -$$
$$m_2 l_1 \ddot{\theta}_1 \sin(\theta_1 - \theta_2) + m_2 g \sin \theta_2 = 0$$

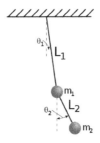

Drawing by Jabber Wok (Creative Commons)

4.2   Partial Differential Equations

Laws of conservation of mass, energy, momentum, and other magnitudes such as electric charge, probability, stress and strain, are normally described with partial differential equations (PDE). What makes PDE special is that the solution depends on space, and possibly on time, i.e. multiple independent variables.

There are elliptical equations (e.g. Laplace and Poisson equations), parabolic equations (heat equation) and hyperbolic equations (wave equation), and they are prescribed by contour conditions of the Dirichlet or Newmann type. The solution methods are different, and in addition, they depend on the type of problem, the geometry, and whether it is a dynamic or static system. Except in very simple geometries, there is usually no simple analytical solution, and it is necessary to use numerical methods based on finite differences, finite elements, and contour volumes.

In this book only two examples of the first method is given, since the other ones are beyond the scope of this book due to their complexity.

4.2.1 Finite Differences

Laplace equation in square geometry

Let us illustrate the method with a simple exercise. We want to solve the Laplace equation in a squared geometry which is 16x16 cm. The boundary conditions are: 20°C along west and south borders, 250 °C along north, and no heat transfer (adiabatic) at the east border.

$$\frac{\partial^2 T}{\partial x^2} + \frac{\partial^2 T}{\partial y^2} = 0.$$

After replacing space derivatives by finite differences, the equation above becomes:

$$\frac{T_{i+1,j} - 2T_{i,j} + T_{i-1,j}}{(\Delta x)^2} + \frac{T_{i+1,j} - 2T_{i,j} + T_{i-1,j}}{(\Delta y)^2} = 0$$

Here the central scheme is used, it can be generalized to any differentiation scheme.

```
#   Partial Differential Equations - Example

from numpy import meshgrid, arange, full, linalg

#   Build a mesh

X, Y = meshgrid(arange(0, 50), arange(0, 50))

#   Initialize

U0 = 30 # first guess
```

```
U=full((50, 50), U0, dtype = float)
V=full((50, 50), U0, dtype = float)

#   Boundary contitions

Unorth = 80
Usouth = 20
Uwest  = 20
Ueast  = 0

U[49,:]   =   Unorth   # Dirichlet
U[0,:]    =   Usouth   # Dirichlet
U[:, 0]   =   Uwest    # Dirichlet
U[:, 48:] =   Ueast    # Neumann

#   Iteration settings

ii=0
dxy = 1
imax = 10000

#   Iterate

while ii < imax:
    #
    #  Calculate central point

    for i in range(1, 49, dxy):
        for j in range(1, 49, dxy):
            U[i, j] = (U[i-1,j]
            + U[i+1,j]
            + U[i,j-1]
            + U[i,j+1])/4
    ii+=1
    error = (linalg.norm(U) -linalg.norm(V))\
            /linalg.norm(U)

    for i in range(0,50):
        for j in range(0,50): V[i][j]=U[i][j]

    if error < 0.0001: ii=imax
```

```
#    Contour plot of the solution U(x,y)

import matplotlib.pyplot as plt
plt.contourf(X, Y, U, 25, cmap = plt.cm.jet)
plt.colorbar()
plt.show()
```

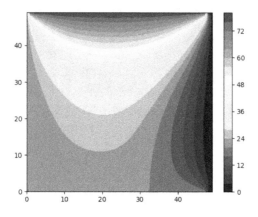

Temperature distribution in the plate.

"Out of the Box"

Time independent Schrodinger equation is the following one:

$$-\frac{\hbar}{2m}\frac{d^2\psi}{dx^2} + v \cdot \psi = E \cdot \psi$$

where ψ is the wave function, x is the spatial dimension, m is the mass of the particle, E is the kinetic energy, V is the height of the potential barrier, and $\hbar$ is the constant of the Plate. This equation is the basis for analyzing the stationary states of atomic systems, and in certain circumstances, there is only a solution for specific (quantified) energy states. The direct analytical solution is not trivial, and one must employ numerical methods that allow approximate solutions to a variety of scenarios. In this example we use the finite difference approach.

I insist upon the view that all is waves. (E. Schrodinger)

If we replace x by x_j, ψ by ψ_j, v by V_j, E by E_j, and $\frac{d^2\psi}{dx^2}$ by $\frac{\psi_{j+1}-2\psi_j+\psi_{j-1}}{\Delta x^2}$, we obtain:

$$\psi_{j+1} = \left[2 - \Delta x^2\frac{2m}{\hbar}(E_j - v_j)\right]\psi_j - \psi_{j-1}$$

Let us write this equation in more comfortable units:

$$\psi_{j+1} = \left[2 - \Delta x^2\frac{2m}{\hbar}1.362 \cdot 10^{-28}(E_j - V_j)\right]\psi_j - \psi_{j-1}$$

where m is the mass of the particle in uma, E and V energies are given in electron-volt, and x is the distance in Armstrong.

Now, we assume that an electron is inside a box with a potential barrier of 20 electron-volt between 2 and 3 Armstrong, and zero before and after the barrier. We are going to find one or more energy levels for this electron inside the box and the associated wave function. The boundary conditions are: $\frac{d\psi}{dx} = 0$, and $\psi = 0$ in a remote point, say $x = 6$ A.

In general terms the procedure that we are going to follow is the following:

1. First, we generate a large number of nodes, say $n = 601$

2. Assume a guess value for energy E, and $\psi_0 = \psi_1 = E$, that is, the same value in order to force the zero gradient condition. This E guess value is irrelevant, because we will normalize the wave function later, but it is necessary to start the iteration.

3. Calculate V_j for all nodes

4. Calculate ψ_j from $j = 2$ to n

5. After the recursive calculation we will see that that ψ_n at the last node will be not zero. Now, we change E guess manually or with an automatic iterative process until you get $\psi_n = 0$.

6. We normalize the wave function:

$$\int_{-\infty}^{+\infty} |\psi(x)|^2 = 1$$

7. Finally, we plot the probability function.

```
# Schrodinger Equation

from numpy import *
from scipy.optimize import fsolve

global f

# Data

m = 5.48e-4        # electron mass
h2 = (6.58e-16)**2  # Square of Plank constant
E0 = 1
dx = 0.01

# Set the x axis

x = arange(0, 6, dx)

# Set the Potential barrier

def V(x):
    if (x > 2 and x < 3):
        y=20
    else:
        y=0
    return y

# Function to generate the wave function
# It returns the wave function and error at x=6

def psiend(p):
    E = p
    global fa
    f = [E] ; f.append(E)
    for i in range(1, len(x)-1):
        f[i] = (2 - (dx**2)*(2*m/h2)*1.362e-28*
            (E-V(x[i])))*f[i-1]-f[i-2]
        f.append(f[i])
    fa = asarray(f)
    error = f[599]
    return error
```

```
#    Iterate E until psi(x=6)=0

E = fsolve(psiend,(E0))
print("E=", E)

# Normalize the psi function and plot

fa = fa/(sum(fa*fa))**0.5
import matplotlib.pyplot as plt
pdf = fa*fa
fig = plt.figure(1)
plt.plot(x, pdf)

# Plot the wave function

fig = plt.figure(2)
plt.plot(x, fa, [0, 2, 2.001, 3, 3.001, 6], [0, 0,
    max(fa), max(fa), 0, 0])
plt.title('Wave function $\Psi$')
plt.xlabel('Distance, uma')
```

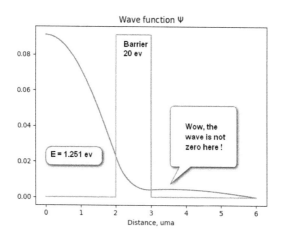

This result is very remarkable: the particle cannot have any energy level, but a specific level. The first level compatible with the Schrodinger equation is E = 1.251 ev; it is a quantum level. In addition, this particle inside the box and with an energy of just 1.251 ev, can surprisingly overcome a much higher barrier of 20 ev, i.e. it has a non-zero probability of crossing this barrier. This result is typical in quantum mechanics, but impossible in classical mechanics and contrary to human intuition. Finally, it is very cool that if the guess energy is progressively increased in the script, different calculated quantum levels emerge (2.5 ev, 9.58 ev, 11.22 ev, 21.08 ev, ...) and the calculated wave functions are progressively more complex than the fundamental wave. But all of them are possible!.

4.2.2   Finite Elements and Finite Volumes

The finite difference method described above is more academic than practical. There are formulations in finite differences that manage to model non-regular geometries although they are somewhat tricky. Currently, the advanced numerical methods used in engineering are the Finite Element Method (FEM) and the Finite Volume Method (FVM). Although both methods are applicable to any system of differential equations in multiphysical partial derivatives, the use of FEM is more consolidated for problems of structural analysis, elastic field, and heat transfer, while FVM takes advantage of the Gauss theorem to simplify the integral equations when it is possible to apply them, and is more dedicated to fluid mechanics (Navier Stokes). There are formulations of both types for Maxwell equations, for gravitational field equations, and other mixed situations.

A detailed description of these methods is out of the scope of this introductory book, and the reader is invited to explore the capabilities of Python applications that use FEM and FVM formulations. Some of these are given below:

- **FEniCS.** The FEniCS Project Version 1.5 M. S. Alnaes, J. Blechta, J. Hake, A. Johansson, B. Kehlet, A. Logg, C. Richardson, J. Ring, M. E. Rognes and G. N. Wells Archive of Numerical Software, vol. 3, 2015, [DOI] Automated Solution of Differential Equations by the Finite Element Method A. Logg, K.-A. Mardal, G. N. Wells et al. Springer, 2012. See

  ```
  https://fenicsproject.org/
  ```

- **SfePy**. R. Cimrman. SfePy - write your own FE application. In P. de Buyl and N. Varoquaux, editors, Proceedings of the 6th European Con- ference on Python in Science (EuroSciPy 2013), pages 65–70, 2014.

  ```
  http://arxiv.org/abs/1404.6391.
  ```

- **FiPy.** J. E. Guyer, D. Wheeler and J. A. Warren, "FiPy: Partial Differential Equations with Python," Computing in Science and Engineering 11 (3) pp. 6-15 (2009). See

  ```
  https://www.nist.gov/publications/
  finite-volume-pde-solver-using-python-fipy
  ```

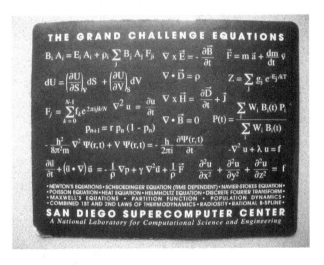

Science is global. [...] Science is a beautiful gift to humanity, we should not distort it. Science does not differentiate between multiple races. (Abdul Kalam)

*When you have eliminated all the impossible, then whatever re-
mains, however improbable, must be the truth* (Sherlock Holmes,
in Arthur Conan Doyle´s tale Sign of the Four)

Chapter 5

Data Science

Data Science is a recent scientific discipline that combines
classical methods such as statistics, regression, data cap-
ture from different sources, and
their treatment to obtain non-
obvious or hidden information
(data mining). Sometimes
sources are incomplete and must be refined (data munging), other
times they are massive (big data) and we must resort to unconven-
tional techniques. In this section we will delve into these concepts
assembled on robust concepts such as estimation, adjustments, and
data mining, and others more typical of artificial intelligence such
as automatic learning, clustering and genetic algorithms. We will
see these concepts approaching the problems in a simple way. Although there are many public and commercial software packages for
this area of knowledge, Python is consolidating itself as a reference
language, almost necessary for the data researcher.

5.1 Univariate Methods - Statistics

Whenever we perform an experiment in which we observe some quantity (strength, number of defects, weight of people, etc.), it is associated with a random variable *xdata*. In Statistics we draw conclusions about the population from the properties of the samples. We do this by calculating point estimates, confidence intervals, upper and lower limits, inference of an adjustment distribution, and other relevant parameters.

In these lines, we will limit ourselves to tiptoe around some essential aspects in this matter. For advanced use, the Python libraries provide a very complete set of functions, tests, and routines to deal with almost any need of the statistics researcher.

5.1.1 Mean and Standard Deviation

Mean and variance of a sample are defined as

$$\bar{x} = \frac{1}{n} \sum_{j=1}^{n} x_j$$

$$s^2 = \frac{1}{n-1} \sum_{j=1}^{n} (x_j - \bar{x})^2$$

Both sample mean $\bar{x}$, and sample variance s^2, are *point estimates* of the population mean μ, and population variance σ^2.

In the attached example both magnitudes are calculated with a very small data set but the method is identical for large volumes of information.

```
#    Sample Statistics

from statistics import *

#    Sample data
xdata = [10.7, 10.9,  8.6,  8.2, 11.8,  9.9,  6.6,
          9.8, 10.1, 10.1,  9.8, 12.3,  8.5, 12.1,
         11.0,  9.2,  9.2,  7.1, 11.8,  6.0, 14.2,
         10.0, 13.1,  9.0, 11.6,  9.6,  8.5, 10.2,
          9.1,  9.3,  7.5, 10.2, 10.2,  9.6,  8.0,
         11.8,  9.7,  7.3,  8.2, 10.0, 10.5, 11.1,
          9.1, 10.2,  7.5, 11.3,  9.8,  7.9,  8.4,
          6.8 , 7.0, 12.4,  8.9,  7.0,  9.8, 10.4,
         10.0,  9.3, 10.0]

# Calculate point estimators

n = len(xdata)
x_sample = mean(xdata)
s_sample = stdev(xdata)

# Print and plot results

print('Results are: ')
print('x_sample =', x_sample)
print('s_sample =', s_sample)
print('n =',n)
#
import matplotlib.pyplot as plt
plt.hist(xdata, color = 'skyblue', edgecolor='blue')
plt.title("Histogram of xdata")
plt.xlabel("Value")
plt.ylabel("Frequency")
```

```
Results are:
x_sample = 9.63050847457627
s_sample = 1.6984023158989985
n = 59
```

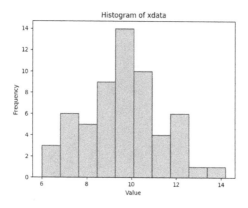

5.1.2 Confidence Limits of μ and σ^2

In general, knowledge of point estimates of population mean and variance is not sufficient, and you want to estimate each parameter within a range that covers a high percentage of the possible values with a high degree of confidence. These are the *Confidence Limits*. Assuming our data comes from a Normal distribution,

$$\bar{x} - t_{\frac{\alpha}{2}} \frac{s}{\sqrt{n}} \leq \mu \leq \bar{x} + t_{\frac{\alpha}{2}} \frac{s}{\sqrt{n}}$$

$$\frac{(n-1)s^2}{\chi_{\frac{\alpha}{2}}^2} \leq \sigma^2 \leq \frac{(n-1)s^2}{\chi_{1-\frac{\alpha}{2}}^2}$$

where t and χ^2, are the Student and Chi2 distributions, and α is the *significance level* (probability of taking a bad decision), typically 5 %.

```
# ... continuation

from scipy.stats import t, chi2
from numpy import sqrt

# significance level

alpha = 0.05

# Confidence intervals for the pop. mean

tstudent = t.ppf(1-alpha/2, n-1)
mu_max = x_sample + tstudent*s_sample/sqrt(n)
mu_min = x_sample - tstudent*s_sample/sqrt(n)

# Confidence interval for the population std dev

Smin = sqrt((n-1)*s_sample**2 / \
            chi2.ppf(1-alpha/2, n-1))
Smax = sqrt((n-1)*s_sample**2 / \
            chi2.ppf(alpha/2, n-1))

print('Confidence Intervals are:')
print(mu_min,' <= mu <=', mu_max)
print(Smin,' <= Sigma2 <=', Smax)
print('\nwith significance level alpha=', alpha)
print('Confidence level gamma = 1-alpha =', 1-alpha)
```

```
Confidence Intervals are:

9.18790242067  <= mu <= 10.0731145285

1.43775470204  <= Sigma2 <= 2.07536904664

with significance level alpha= 0.05
Confidence level gamma = 1-alpha = 0.95
```

5.1.3 Population Bounds

Another very useful quantity in statistical inference is the value that covers a given percentage (say 95%) of all possible population values. This is the so-called *upper bound*.

If you knew the exact values of μ and σ of the population, it would suffice to calculate the upper bound as $x_{95\%} = \mu + 1.645\sigma$. But these values are unknown. Instead, starting from the known sample values, $\bar{x}$ and s, the upper bound value can also be estimated with Owen's distribution,

$$x_{95\%} \simeq \bar{x} + k_{95\%} \cdot s$$

```
# ... continuation

# Estimation of the 95/95 upper bound of population

owen2 = lambda n: 1.96 + 2.1758/sqrt(n) + 5.7423/n
owen1 = lambda n: 1.6449 + 2.4417/sqrt(n) + 3.8171/n

up95_1tail = x_sample + owen1(n)*s_sample
up95_2tail = x_sample + owen2(n)*s_sample

print('Upper bounds of the population are:')
print('Up95_1tail =',up95_1tail,': 5% at right')
print('Up95_2tail =',up95_2tail,
': 2.5% at right and left')
```

```
Upper bounds of the population are:
Up95_1tail = 13.07398328 : 5% at right
Up95_2tail = 13.60577560 : 2.5% at right and left
```

5.1.4 Test for a Distribution

Given a sample $[x_1, ..., x_n]$, we want to test the hypothesis that a given function F(x) is the distribution function from which the sample was taken. There are a number of different tests to verify the hypothesis. Currently, the recommended are Shapiro-Wilk (just to check normality) and the Anderson-Darling tests, both described below.

Shapiro-Wilk Normality Test

```
# ... continuation

# Shapiro-Wilk Normality test

from scipy.stats import shapiro

sw = shapiro(xdata)
print('Shapiro-Wilks Test p=',sw[1],' W=',sw[0])
print('If p < w, do not reject Normality')
```

```
Shapiro-Wilks Test
p= 0.7642765641212463  W= 0.9866541028022766
If p < w, do not reject Normality
```

In this case, we do not reject the hypothesis that the sample comes from a normal distribution.

Anderson-Darling Test

```
# ...continuation

# Anderson Darling test

from scipy.stats import anderson

AD=anderson(xdata, dist='norm')

print(AD)
```

```
AndersonResult(statistic = 0.3259179212011,

critical_values =
    array([0.543, 0.619, 0.742, 0.866, 1.03]),

significance_level =
    array([ 15.,   10.,    5.,   2.5, 1.]))
```

Here, since AD = 0.3259 is lower than critical value 0.742 at a 5% significance level, the hypothesis of normality is not rejected.

5.2 Bivariate Methods

5.2.1 Least Squares

The least-squares (LS) method has many applications such as smoothing and correcting experimental data, performing interpolations, and the application we are interested in in this section, establishing mathematical relationships between two variables. The result is a polynomial that fits the data consistently, and that can be used instead of the data itself as a formula, although it does not necessarily go through all of them. It is a tool widely used in many fields of engineering.

The first formulations of the least-squares method date back to 1805 (Legendre and Gauss), when the predictive power of the method was shown in situations where observational data incorporate some degree of error.

Our goal is to find a polynomial $P_m = \sum_{i=0}^{m} a_i x^i$ of lesser degree than the pure polynomial, which passes near the points with non-oscillating behavior. The coefficients $a_0, ...a_n$ are determined in such a way as to minimize the quadratic error of the fit at all points. In matrix form:

$$\begin{bmatrix} n+1 & \sum x_i & \sum x_i^2 & ... & \sum x_i^m \\ \sum x_i & \sum x_i^2 & ... & ... & \sum x_i^{m+1} \\ \sum x_i^2 & ... & ... & ... & \sum x_i^{m+2} \\ ... & & & & \\ \sum x_i^m & \sum x_i^{m+1} & ... & ... & \sum x_i^{2m} \end{bmatrix} \begin{Bmatrix} a_0 \\ a_1 \\ \vdots \\ a_m \end{Bmatrix} = \begin{Bmatrix} \sum y_i \\ \sum x_i y_i \\ \vdots \\ \sum x_i^m y_i \end{Bmatrix}$$

and solve for a.

Normally we have many n+1 points and look for a polynomial of lesser degree m<n. The higher the grade of the adjustment polynomial, the closer it gets to the real data points at which it is executed. But keep in mind that we are not looking for a pure interpolation polynomial, but a smoother, lesser order polynomial that does not oscillate, and that somehow compensates for inherent data errors.

As a suggestion, you can represent the dots on a scatter plot. Next, you can test a degree that is equal to the number of trend changes plus one, and then increase the degree of adjustment of LS. A plot with all the cases can be of great help. Many authors indicate that the best adjustment is the one that results in a higher regression coefficient. However, generalization of this criterion leads to a pure interpolation polynomial, which by definition will give the maximum adjustment of 100% through all points, but that is not what we are looking for. Instead, it is more advisable to gradually increase the degree of the LS regression polynomial, and graph the result to observe behavior, not only at the data but also in the intermediate intervals. In the following example, the polynomial adjustment was increased to an eighth degree, but beyond that, the polynomial began to oscillate. It depends entirely on the data set.

```
#    Least Squares
#

from numpy import array, poly1d, polyfit, arange
from scipy.interpolate import interp1d

#    Data

xdata = array([ 2,   3,   4,   6, 12, 18, 22, 33, 40,
               45, 50, 57])
ydata = array([4.5, 10, 16, 37, 120, 100, 83.9,
               65, 64, 66, 70, 71])

#    Calculate all LS fits

a2 = polyfit(xdata, ydata, 2)
a3 = polyfit(xdata, ydata, 3)
a8 = polyfit(xdata, ydata, 8)

#    Finer x range just for plotting

xvals = arange(2,57, 0.5)
yvals_LST2 = poly1d(a2) (xvals)
yvals_LST3 = poly1d(a3) (xvals)
yvals_LST8 = poly1d(a8) (xvals)

#    Plot LS fits

import matplotlib.pyplot as plt
plt.close('all') # erase old plots
fig = plt.figure(1)
plt.plot(xvals, yvals_LST2, 'b-',
```

```
        xvals, yvals_LST3, 'r-',
        xvals, yvals_LST8, 'y-',
        xdata, ydata, 'bo')
plt.legend(('LS2','LS3','LS8','data'))

#   Print results

print('Coefficients are:',a8 )
```

```
Coefficients are: [ 1.21242621e-09  -3.31174174e-07
                    3.75413629e-05  -2.27553315e-03
                    7.89102201e-02  -1.54036561e+00
                    1.51737533e+01  -5.36623265e+01
                    6.55567161e+01]
```

The array is ordered from highest to lowest degree term ($a_n, ... a_0$).

5.2.2   Correlate two Variables with Multiple Choices

In some occasions, after representing the points with a scatter diagram, we perceive that there can be alternative regressions or more suitable than a simple polynomial. Perhaps we want to explore an exponential relationship, or logarithmic, or potential, or other types of functions. In the attached script this type of exploration is performed on a very small set of data pairs.

```
#   Fit many equations

from numpy import *
import matplotlib.pyplot as plt
from scipy.optimize import curve_fit

# Data to fit

xd = array([2, 5, 7, 12, 13, 18, 33, 50, 66, 72])
```

```
yd = array([5, 23, 50, 130, 135, 250, 1000,
            1300, 2010, 1800])

# Define functions

def func1(x, a, b):      return a*x + b
def func2(x, a, b, c):   return a*x**b + c
def func3(x, a, b, c):   return a*x**2 + b*x + c
def func4(x, a, b, c):   return a*exp(-b*x) + c
def func5(x, a, b):      return a*log(x) + b
def func6(x, a, b):      return 1/(a + b*x)
def func7(x, a, b):      return 1/(a + b*log(x))
def func8(x, a, b):      return exp(a + b*log(x))

# Fit all equations

p1, p = curve_fit(func1, xd, yd)
p2, p = curve_fit(func2, xd, yd)
p3, p = curve_fit(func3, xd, yd)
p4, p = curve_fit(func4, xd, yd)
p5, p = curve_fit(func5, xd, yd)
p6, p = curve_fit(func6, xd, yd)
p7, p = curve_fit(func7, xd, yd)
p8, p = curve_fit(func8, xd, yd)


#   Some statistics (f)its, s(igma), (r)egression
    coef.

f1 = func1(xd, *p1) ; s1 = std(yd-f1)
r1 = corrcoef(f1, yd)[1,0]

f2 = func2(xd, *p2) ; s2 = std(yd-f2)
r2 = corrcoef(f2, yd)[1,0]

f3 = func3(xd, *p3) ; s3 = std(yd-f3)
r3 = corrcoef(f3, yd)[1,0]

f4 = func4(xd, *p4) ; s4 = std(yd-f4)
r4 = corrcoef(f4, yd)[1,0]
```

```
f5 = func5(xd, *p5) ; s5 = std(yd-f5)
r5 = corrcoef(f5, yd)[1,0]

f6 = func6(xd, *p6) ; s6 = std(yd-f6)
r6 = corrcoef(f6, yd)[1,0]

f7 = func6(xd, *p7) ; s7 = std(yd-f6)
r7 = corrcoef(f7, yd)[1,0]

f8 = func6(xd, *p8) ; s8 = std(yd-f6)
r8 = corrcoef(f8, yd)[1,0]

# Print results

print('\n', 1,p1, 'rho ', r1,'sigma ', s1)
print('\n', 2,p2, 'rho ', r2,'sigma ', s2)
print('\n', 3,p3, 'rho ', r3,'sigma ', s3)
print('\n', 4,p4, 'rho ', r4,'sigma ', s4)
print('\n', 5,p5, 'rho ', r5,'sigma ', s5)
print('\n', 6,p6, 'rho ', r6,'sigma ', s6)
print('\n', 7,p7, 'rho ', r7,'sigma ', s7)
print('\n', 8,p8, 'rho ', r8,'sigma ', s8)
```

```
1 [  29.7937406   -157.96598862]
  rho  0.986540071992 sigma  122.123676816

2 [  26.16754764    1.02930359 -141.0862143 ]
  rho  0.986584107494 sigma  121.925094571

3 [ -2.98160679e-02   3.19732973e+01   -1.77280301e+02]
  rho  0.986669046694 sigma  121.541111603

4 [ -94.35437572   52.19799414   670.29999874]
  rho  nan sigma  746.843899352

5 [ 589.75840766 -984.77289774]
  rho  0.885746319527 sigma  346.662266109

6 [  2.73843105e-03   -3.14078488e-05]
  rho  0.892645787896 sigma  353.924040304

7 [ 0.00585923 -0.00125729]
```

```
   rho   0.305853230166  sigma   353.924040304

8 [ 2.34910938   1.23021951]
   rho   -0.690755869662  sigma   353.924040304
```

In this case, observing the correlation coefficients and the standard deviation of each regression, we conclude that the first three are adequate, and the remaining ones are not. Taking for example the second one, we have:

$$y = 26.167x^{1.0293} - 141.08, \quad \rho = 98.6\%$$

5.3   Multivariate Methods

In scientific or industrial research environments, there is an occasional need to find expressions that relate several inputs to one or more outputs. These inputs and outputs, in original form or transformed with simple functions, may be adjusted to multiple linear regressions with maximum likelihood or least squares methods. The reader will notice that some of the concepts in this section have already been introduced previously. The example that we are going to follow is very simple, both in the number of variables (fields) and in the number of records, although it will serve to expose a practical technique of multiple regression.

5.3.1   Read and Describe Data

The example data source is an Excel file with only 18 rows but the method is applicable to a much larger size, perhaps several thousand records.

To read the Excel file, we use the Pandas library, widely used in Python applications for Data Sciences. With it we create a Python object called dataframe, which can then be interrogated with the commands keys, describe, head, and dtypes, to have clear information of what is in each of the fields.

```python
# Multivariate Regression

from numpy import corrcoef, transpose, around
import statsmodels.api as sm
import matplotlib.pyplot as plt
import pandas as pd
import os

os.system("cls") # clear screen

# (1) Read all data

mydata = pd.read_excel("Data_prod.xlsx", "Sheet1")
print("\n(1) Read data")

# (2) Description

print("\n(2) Fields description")
print('keys are \n',   mydata.keys())
print('Describe \n',   mydata.describe())
print('Head is \n',   mydata.head())
print('Types are \n', mydata.dtypes)
```

```
(1) Read data

(2) Fields description
keys are
Index(['Coal', 'Iron', 'Wood', 'Nickel', 'Gas', 'Prod'], dtype='object')
Describe
            Coal       Iron       Wood     Nickel        Gas        Prod
count  18.000000  18.000000  18.000000  18.000000  18.000000  18.000000
mean    9.723501  34.495216  19.509280   4.513004 203.454656  36.183973
std     3.137497   6.254158   3.590165   1.642908  59.972480  41.724941
min     5.159120  26.455069  12.453450   2.072828 103.645979 -18.966415
25%     7.212349  29.246786  17.009680   3.315510 151.586670   2.333159
50%     9.828290  34.839651  19.529569   4.389162 211.009096  29.259057
75%    11.834043  38.594353  22.731731   5.862938 245.656210  73.524070
```

101

```
max     14.940519   48.794617   23.808247    6.973731  286.441744  105.355848
Head is
             Coal        Iron        Wood      Nickel         Gas        Prod
0        7.445735   29.410907   22.235706    6.154254  228.400468   18.553170
1       12.236518   26.455069   14.346276    4.220295  286.441744   82.790689
2       14.552268   39.016811   19.439719    4.390807  285.368261   85.221445
3       14.289527   27.160074   17.980871    6.670089  135.749267  105.355848
4        7.881015   35.927669   17.568952    2.394150  190.091056   11.448009


Types are
Coal       float64
Iron       float64
Wood       float64
Nickel     float64
Gas        float64
Prod       float64
dtype: object
```

At this point, we have already read the data and we know the names of each of the fields, and we have some elementary statistics (mean, standard deviation, etc.), as well as the type of data (integer, float, or other). If there were any empty positions, then a *munging* method would have to be used as explained in the example in the next section "Data Mining".

The next step is to rename each variable in a short name to handle them comfortably.

```
#   . . .

#   (3) Set shorter names for convenience

print("\n(3) Set shorter names")
c = mydata['Coal']
i = mydata['Iron']
w = mydata['Wood']
n = mydata['Nickel']
g = mydata['Gas']
p = mydata['Prod']
```

5.3.2 Are the data correlated with each other?

This question refers to whether the variations of some variables are explained by the variations of other variables. To answer this question we calculate the correlation matrix, in which each component is the Pearson coefficient that can be used to measure the degree of relationship of two variables as long as both are quantitative and continuous.

```
#    ...

#  (4) Correlation R-coeff (Pearson) to ALL
     candidates

YDEP    = p
XALL    = [c, i, w, n, g, c*c, i*i, c*w*n]

matrix = corrcoef(YDEP, XALL)
print('\n(4) Correlation matrix (Pearson) of YDEP vs
    . XALL:\n', around(matrix, 2))
```

```
(4) Correlation matrix (Pearson) of YDEP vs. XALL:
[[1.     0.96 -0.46 -0.06  0.09  0.36  0.95 -0.45  0.59]
 [ 0.96  1.   -0.18 -0.1   0.15  0.32  0.99 -0.16  0.64]
 [-0.46 -0.18  1.   -0.09  0.16 -0.24 -0.2   1.   -0.03]
 [-0.06 -0.1  -0.09  1.    0.04  0.03 -0.11 -0.12  0.25]
 [ 0.09  0.15  0.16  0.04  1.   -0.07  0.16  0.17  0.78]
 [ 0.36  0.32 -0.24  0.03 -0.07  1.    0.32 -0.26  0.12]
 [ 0.95  0.99 -0.2  -0.11  0.16  0.32  1.   -0.18  0.64]
 [-0.45 -0.16  1.   -0.12  0.17 -0.26 -0.18  1.   -0.04]
 [-0.09 -0.12 -0.06  1.    0.04  0.01 -0.14 -0.09  0.22]
 [ 0.59  0.64 -0.03  0.25  0.78  0.12  0.64 -0.04  1.   ]]
```

5.3.3 Regression

(1) The correlation matrix calculated in the previous step gives us a clue as to which are the strongest candidates to enter the regression. In the example we see that the highest correlations are obtained with c and with c^2, both with a similar ρ value. In general, all things being equal, we prefer the simplest form and therefore we introduce XCANDIDATES = [c] as the first trial, and then run the script below.

(2) After executing the first trial we see that residuals have zero correlation with c but instead they have a high degree of correlation ($r = 0.98$) with i. This indicates that we can include this variable into the candidates, XCANDIDATES = [c, i].

(3) A new execution results in a correlation matrix whose terms are all quite reduced, indicating that it is no longer another candidate for regression, and we are done.

```
#    ...

#  (5) Try a regression
#
#     Add or remove vars into XCANDIDATES array
#     until residuals show no significant
#     correlation with remaining vars.

XCANDIDATES = [c, i] # Candidates. First run is []
intercept   = 1        # 1 or 0

# --- No changes below this line!

if XCANDIDATES != []:
    if intercept==1:
        stat = sm.OLS(YDEP, sm.add_constant(
                transpose(XCANDIDATES))).fit()
```

```
    else:
        stat = sm.OLS(YDEP,(transpose(
            XCANDIDATES))).fit()

    # Print and plot relevants

    plt.figure(1)
    plt.plot(YDEP,ypred,'o',[min(YDEP),max(YDEP)],[
    min(YDEP),max(YDEP)],'-')
    plt.xlabel('y read')
    plt.ylabel('y predicted')

    plt.figure(2)
    plt.plot(YDEP,rsid,'*')

    matrixr = corrcoef(rsid, XALL)
    print('\n(5) Correlation matrix (Pearson) of
    RESIDUALS vs XALL:\n',around(matrixr,2))
    print('\n(6) Regression of YDEP vs. XCANDIDATES\
    n',stat.summary())

    # option print estimates and residuals

    print('\n(7) Predicted values:\n', ypred)
    print('\n(8) Residuals:\n',rsid)
```

```
(5) Correlation matrix (Pearson) of RESIDUALS vs XALL:
[[ 1.    -0.     0.98 -0.12  0.2  -0.19 -0.02  0.98 -0.09  0.08]
 [-0.     1.    -0.18 -0.1   0.15  0.32  0.99 -0.16 -0.12  0.64]
 [ 0.98 -0.18  1.    -0.09  0.16 -0.24 -0.2   1.    -0.06 -0.03]
 [-0.12 -0.1  -0.09  1.     0.04  0.03 -0.11 -0.12  1.     0.25]
 [ 0.2   0.15  0.16  0.04  1.    -0.07  0.16  0.17  0.04  0.78]
 [-0.19  0.32 -0.24  0.03 -0.07  1.     0.32 -0.26  0.01  0.12]
 [-0.02  0.99 -0.2  -0.11  0.16  0.32  1.    -0.18 -0.14  0.64]
 [ 0.98 -0.16  1.    -0.12  0.17 -0.26 -0.18  1.    -0.09 -0.04]
 [-0.09 -0.12 -0.06  1.     0.04  0.01 -0.14 -0.09  1.     0.22]
 [ 0.08  0.64 -0.03  0.25  0.78  0.12  0.64 -0.04  0.22  1.   ]]

(6) Regression of YDEP vs. XCANDIDATES
OLS Regression Results
==========================================================================
Dep. Variable:              Prod   R-squared:                      0.912
Model:                       OLS   Adj. R-squared:                 0.907
Method:            Least Squares   F-statistic:                    166.5
Date:           Thu, 11 Jul 2019   Prob (F-statistic):          7.13e-10
Time:                   10:53:24   Log-Likelihood:               -70.278
No. Observations:             18   AIC:                            144.6
```

```
Df Residuals:                    16    BIC:                         146.3
Df Model:                         1
Covariance Type:           nonrobust
==============================================================================
                 coef      std err          t      P>|t|    [95.0% Conf. Int.]
------------------------------------------------------------------------------
const         -87.3293      10.031      -8.706      0.000    -108.594  -66.065
x1             12.7026       0.984      12.904      0.000      10.616   14.789
==============================================================================
Omnibus:                        1.416    Durbin-Watson:               2.135
Prob(Omnibus):                  0.493    Jarque-Bera (JB):            1.099
Skew:                          -0.564    Prob(JB):                    0.577
Kurtosis:                       2.563    Cond. No.                    34.4
==============================================================================

Warnings:
[1] Standard Errors assume that the covariance matrix of the errors is co
rrectly specified.

(7) Predicted values:
[    7.25052937    68.10571286    97.52165431    94.1841668     12.77969848
  -12.41694769   -21.79532186   102.45342166    35.30926689    58.85037781
   34.4107002     39.72086698    46.17727223    -2.20046537     3.29772748
   44.60328454    64.37420495   -21.31463313]

(8) Residuals:
0      -11.302640
1      -14.684976
2       12.300209
3      -11.171681
4        1.331690
5      -13.362864
6       -2.828906
7        4.939448
8       28.814378
9       -7.222617
10       5.802361
11       9.811093
12     -14.599349
13      10.831047
14      14.251860
15       5.689690
16     -11.633557
17      -6.965186
Name: Prod, dtype: float64
```

(4) Now is the time to look at the P and t values associated with each coefficient. The P value tells you how confident you can be that each individual variable has some correlation with the dependent variable. The t statistic is the coefficient divided by its standard error. It can be thought of as a measure of the precision with which the regression coefficient is measured. If a coefficient is large compared to its standard error, then it is probably different from 0.

Since we have obtained $r^2 = 1.00$ and also the P value of each coefficient is much less than 0.05, which is the cut-off value to accept or reject a coefficient, our result is very satisfactory. Therefore, the final regression becomes:

$$P = -11.0956 + 11.984c - 2.0074i \simeq -11 + 12c - 2i$$

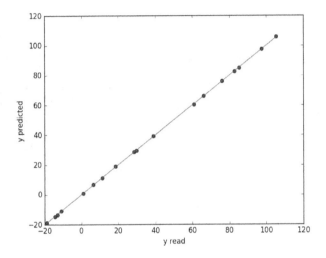

The calculated regression predicts Y values very close to the real ones. A representation of this type is the one that best visualizes if there are potential local deviations that need to be investigated.

5.4 Data Mining

Aristotle in antiquity, and recently Francis Bacon coined the idea that information is power. It is well known that the amount of information stored by humanity is growing at an exponential rate, with a doubling rate of approximately 15 years. We live in a historical moment in which there is much more information than we can digest, with the added difficulty of discerning what is significant and what is not.

The general problem of data mining is to extract that valuable hidden information, find the truth from a set of data with different structures and origins, establish relationships between parameters, fill gaps, discriminate significant variables, and discover undeclared categories. It is a

Truth is what it is and it's still true even if you think the other way around. (Antonio Machado)

highly demanded branch of data science in industry and science, which is constantly evolving.

The following exercise is a small demonstration of some of the techniques currently being used for data mining with Python.

5.4.1 Reading File and Initial Description

The basic structure for data mining is the creation of a dataframe with the Pandas library, because the information is stored with a comfortable structure for further treatment. In this exercise we import a file in *csv* format.

Once we have read the file with the raw data we generate a description of each field.

```python
# Data Mining

import pandas as pd
import matplotlib.pyplot as plt
import statsmodels.api as sm
from numpy import transpose, around, corrcoef
import os ; os.system("cls") # clear console


# (1) Import the csv file into a dataframe

dfraw = pd.read_csv('Data_mining_example_voids.csv',
    sep=";", header=0)
print("\n(1) Read data")

# (2) Describe fields

print('\n(2) Describe fields of raw file')
print(dfraw.describe())
print('Are there empty records? \n', dfraw.isnull().
    any())
```

```
(1) Read data

(2) Describe fields of raw file
         xdata        ydata        zdata        adata      bresult  \
count  5023.000000  5019.000000  5023.000000  5021.000000  5011.000000
mean      5.001061    10.009881    20.023828    16.011884     4.979437
std       1.003738     2.011106     2.840323     8.009033     5.267090
min       1.417394     1.879635    15.000503     2.012465   -18.506657
25%            NaN          NaN          NaN          NaN          NaN
50%            NaN          NaN          NaN          NaN          NaN
75%            NaN          NaN          NaN          NaN          NaN
max       8.816980    17.230994    24.999100    29.997229    27.395848

       cresult
count  5009.000000
mean     54.941322
std      13.188443
min     -12.586850
25%           NaN
```

```
50%              NaN
75%              NaN
max        106.713544
Are there empty records?
xdata            True
ydata            True
zdata            True
adata            True
bresult          True
cresult          True
dtype: bool
```

5.4.2 Data Munging and Final Data Description

The statistics of each field is eloquent: the count of each field is not the same, in addition it is not possible to determine the percentiles, and there are even empty records. To continue with the analysis it is necessary to clean the dataframe in a process known as *munging*. In this example we will eliminate the records in which there is an empty field of type "NaN", although another good possibility could be to fill in the gaps with the average value of the field, or a rolling average, or a clever interpolation, among other possibilities.

Once the Munging is done, we have a complete dataframe. With the instructions of step (4) in script below we obtain a very complete description of each field: its elementary statistics, format, appearance and type, and in step (6) we draw the histogram of each field. On the other hand, we know that the independent parameters or inputs are *xdata, ydata, zdata, adata,* and the dependent or outputs are *bresult* and *cresult*.

```python
# ... continuation

# (3) Clean: remove rows with empty info

print('\n(3) Clean raw file')
dfclean=dfraw.dropna(axis=0,how='any')
```

```
## Option fill with means - uncomment if needed
# dfclean=dfraw
# dfclean.fillna(dfclean.xdata.mean(),inplace=True)


# (4) Provide basic information of clean file

print('\n(4) Information abaout clean file')
print('keys are \n', dfclean.keys())
print('describe \n', dfclean.describe())
print('Head is \n', dfclean.head())
print('types are \n', dfclean.dtypes)


# (5) Set shorter names for convenience

x, y, z = dfclean.xdata, dfclean.ydata, dfclean.
    zdata
a, b, c = dfclean.adata, dfclean.bresult,
          dfclean.cresult
print('\n(5) set shorter names')

# (6) Plot histograms for each field

# index is plots file, col
plt.figure(0)
plt.subplot(221) ; plt.hist(x) ; plt.ylabel('x')
plt.subplot(222) ; plt.hist(y) ; plt.ylabel('y')
plt.subplot(223) ; plt.hist(z) ; plt.ylabel('z')
plt.subplot(224) ; plt.hist(a) ; plt.ylabel('a')
plt.show()

plt.figure(1)
plt.subplot(131) ; plt.hist(x) ; plt.ylabel('b')
plt.subplot(132) ; plt.hist(y) ; plt.ylabel('c')
plt.subplot(133) ; plt.hist(z) ; plt.ylabel('d')
plt.show()
print('\n(6) Plot histograms')
```

```
(3) Clean raw file

(4) Information abaout clean file
keys are
Index(['xdata', 'ydata', 'zdata', 'adata', 'bresult', 'cresult'], dtype='object
    ')
describe
xdata          ydata          zdata          adata          bresult      \
count   4965.000000   4965.000000   4965.000000   4965.000000   4965.000000
mean       5.002651     10.009529     20.023915     16.004759      5.004645
std        1.003833      2.013321      2.839494      8.013139      5.155863
min        1.417394      1.879635     15.000503      2.012465    -13.185724
25%        4.342196      8.655697     17.591869      9.239606      1.423812
50%        4.989011     10.003275     20.039535     15.946807      5.089917
75%        5.680604     11.361209     22.415602     22.957083      8.584430
max        8.816980     17.230994     24.999100     29.997229     24.398357


cresult
count   4965.000000
mean      55.054071
std       12.901013
min        6.865223
25%       46.408883
50%       55.250589
75%       63.699205
max      106.713544
Head is
xdata          ydata      zdata        adata     bresult     cresult
0    4.832916     11.067964   21.567132    7.314386    5.050946   60.031814
1    5.658365      8.809022   15.334497   16.152045    7.184497   54.739681
2    4.694674      6.063163   19.607394   13.449413   -3.057900   30.922837
3    4.567958      9.412653   23.489065    3.634557    0.180001   46.434540
4    5.943290     10.790661   24.931207   14.782276    2.217208   57.754122
types are
xdata        float64
ydata        float64
zdata        float64
adata        float64
bresult      float64
cresult      float64
dtype: object

(5) set shorter names

(6) Plot histograms
```

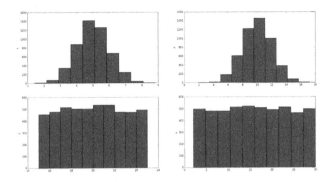

Histograms of x and y (top), z and a (bottom)

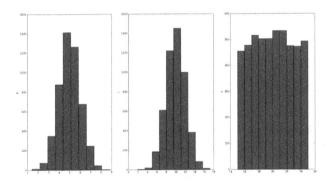

Histograms of b, c, and d

Correlations

Now, in step (7) we find out the extent to which paired variables move together. To do this, we calculate the correlation matrix (Pearson) for each pair of variables. Also, we can represent the graphs of

113

the input - output pairs that will give us a good idea of the dependencies.

```
# ... continuation

# (7) Correlations

print('\n(7) Correlations')
matrix = corrcoef([x, y, z, a, b, c])
print('\nCorrelation matrix (Pearson) of ALL fields
    :\n', around(matrix, 2))

# Plot X - Y relations

plt.figure(4)
plt.subplot(241) ; plt.plot(x, b, '.')
plt.xlabel('x')  ; plt.ylabel('b')
plt.subplot(242) ; plt.plot(y, b, '.')
plt.xlabel('y')  ; plt.ylabel('b')
plt.subplot(243) ; plt.plot(z, b, '.')
plt.xlabel('z')  ; plt.ylabel('b')
plt.subplot(244) ; plt.plot(a, b, '.')
plt.xlabel('a')  ; plt.ylabel('b')

plt.subplot(245) ; plt.plot(x, c, '.')
plt.xlabel('x')  ; plt.ylabel('c')
plt.subplot(246) ; plt.plot(y, c, '.')
plt.xlabel('y')  ; plt.ylabel('c')
plt.subplot(247) ; plt.plot(z, c, '.')
plt.xlabel('z')  ; plt.ylabel('c')
plt.subplot(248) ; plt.plot(a, c, '.')
plt.xlabel('a')  ; plt.ylabel('c')

plt.show()
```

```
(7) Correlations

Correlation matrix (Pearson) of ALL fields:
[[ 1.    0.01 -0.02  0.02  0.21  0.24]
 [ 0.01  1.   -0.03 -0.02  0.8   0.94]
 [-0.02 -0.03  1.    0.02 -0.58 -0.26]
 [ 0.02 -0.02  0.02  1.   -0.02 -0.02]
 [ 0.21  0.8  -0.58 -0.02  1.    0.93]
 [ 0.24  0.94 -0.26 -0.02  0.93  1.  ]]
```

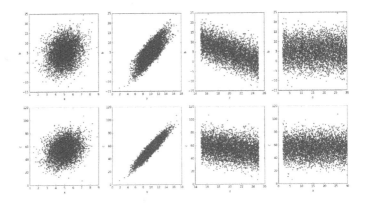

Main cross correlation plots

5.4.3 Regressions

Looking at the correlation matrix and the graphs, we conclude that b and c depend on y and z. In (8) we will look for the regressions using the same library as in the previous example "Regression". The most direct thing is to go testing with several variables systematically until you find a good balance between low number of variables and high R-squared value. Care should be taken not to introduce input variables that are dependent on each other, since this complicates the regression and also introduces a variability that is not necessary. In this example, we reached the dependence that we had assumed after the inspection of the correlation matrix and the graphs: b and c depending on y and z.

```python
# ... continuation

# (8) Regressions

#    Define a regression function
def reg(X,Y,const):
    if const==1:
        stat = sm.OLS(Y, sm.add_constant(transpose
            (X))).fit()
    else:
        stat = sm.OLS(Y, (transpose(X))).fit()
    return stat

print('\n(8) Regressions')

# **** Regression to b

X = [y, z] ; Y = b ; const=1
stat = reg(X, Y, const)
print(stat.summary())
plt.figure(2)
plt.plot(Y, stat.predict(), '.')
plt.ylabel('Ypred vs Y')

# **** Regression to c

X = [y, z] ; Y = c ; const=1
stat = reg(X, Y, const)
print(stat.summary())
plt.figure(3)
plt.plot(Y, stat.predict(), '.')
plt.ylabel('Ypred vs Y')
```

```
(8) Regressions
OLS Regression Results
========================================================================
Dep. Variable:                bresult   R-squared:                  0.949
Model:                            OLS   Adj. R-squared:             0.949
Method:                 Least Squares   F-statistic:            4.574e+04
Date:                Tue, 09 Jul 2019   Prob (F-statistic):          0.00
Time:                        18:55:34   Log-Likelihood:            -7821.7
No. Observations:                4965   AIC:                    1.565e+04
Df Residuals:                    4962   BIC:                    1.567e+04
Df Model:                           2
Covariance Type:            nonrobust
========================================================================
                 coef    std err          t      P>|t|    [0.025   0.975]
------------------------------------------------------------------------
const          5.1307      0.146     35.024      0.000     4.844    5.418
x1             2.0031      0.008    242.784      0.000     1.987    2.019
x2            -1.0076      0.006   -172.240      0.000    -1.019   -0.996
========================================================================
Omnibus:                        1.870   Durbin-Watson:              2.014
Prob(Omnibus):                  0.393   Jarque-Bera (JB):           1.830
Skew:                           0.010   Prob(JB):                   0.401
Kurtosis:                       2.908   Cond. No.                    199.
========================================================================

Warnings:
[1] Standard Errors assume that the covariance matrix of the errors is cor
rectly specified.
OLS Regression Results
========================================================================
Dep. Variable:                cresult   R-squared:                  0.943
Model:                            OLS   Adj. R-squared:             0.943
Method:                 Least Squares   F-statistic:            4.117e+04
Date:                Tue, 09 Jul 2019   Prob (F-statistic):          0.00
Time:                        18:55:35   Log-Likelihood:            -12623.
No. Observations:                4965   AIC:                    2.525e+04
Df Residuals:                    4962   BIC:                    2.527e+04
Df Model:                           2
Covariance Type:            nonrobust
========================================================================
coef          std err          t                P>|t|    [0.025   0.975]
------------------------------------------------------------------------
const         15.3043      0.385     39.723      0.000    14.549   16.060
x1             6.0077      0.022    276.866      0.000     5.965    6.050
x2            -1.0180      0.015    -66.166      0.000    -1.048   -0.988
========================================================================
Omnibus:                        0.303   Durbin-Watson:              2.040
Prob(Omnibus):                  0.859   Jarque-Bera (JB):           0.278
Skew:                           0.017   Prob(JB):                   0.870
Kurtosis:                       3.015   Cond. No.                    199.
========================================================================

Warnings:
[1] Standard Errors assume that the covariance matrix of the errors is
    correctly specified.
```

At the end of the process, we come to the regression lines:

$$b \simeq 5.13 + 2y - z$$
$$c \simeq 15.3 + 6y - z$$

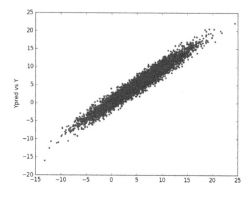

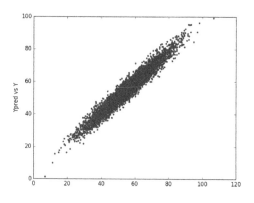

Final regressions

5.4.4 Clustering

Within the scope of Data Science, a very impressive application is the identification of data groupings or clusters. The scikit-learn module is a *machine learning* application that includes classification, regression and group analysis algorithms, among which are support vector machines, random forests, gradient boosting, K-means and DBSCAN.

The example below consists of a cloud of points in which for some reason we believe there are four groups. The Python script is able to categorize each point and also finds the center of each cluster.

```python
# cluster analysis

import numpy as np
import matplotlib.pyplot as plt
import pandas as pd
from sklearn.cluster import KMeans
import os; os.system("cls") # clear screen

# Read and plot data

dfdata = pd.read_excel("locus.xlsx","Hoja1")

x = dfdata['x']
y = dfdata['y']

plt.figure(1)
plt.plot(x, y, 'x')
plt.title('Data before clustering')

# Create 2D list

locus = np.column_stack((x, y))
plt.ylim((-20, 25)); plt.xlim((-20, 15))
```

119

```
# Create kmeans object

kmeans = KMeans(n_clusters = 4)

# Fit kmeans object to data

kmeans.fit(locus)

# Calculate cluster centers

cc = kmeans.cluster_centers_
print("Cluster centers =\n", cc)

# Plot data after clustering

yk = kmeans.fit_predict(locus)

plt.figure(2)
plt.scatter(locus[yk ==0,0], locus[yk == 0,1],
            s=22, c='green')
plt.scatter(locus[yk ==1,0], locus[yk == 1,1],
            s=22, c='red')
plt.scatter(locus[yk ==2,0], locus[yk == 2,1],
            s=22, c='cyan')
plt.scatter(locus[yk ==3,0], locus[yk == 3,1],
            s=22, c='blue')

plt.scatter(cc[:,0],cc[:,1], s=100, marker='s',
            facecolor='yellow', edgecolors='r')
plt.title('Data after clustering')
```

```
Cluster centers =
[[ -5.33398595  -0.82875769]
 [ -2.19396073  10.40955769]
 [  1.70615288  -4.74432972]
 [ -3.48173682  -6.80165435]]
```

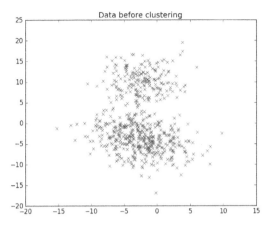

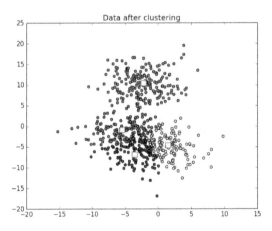

Each point is classified as belonging to a cluster automatically determined by the algorithm. The four categories were hidden and are now represented by a single point, the center of the cluster.

5.5 Genetic Algorithms

One of the most suggestive branches of artificial intelligence is problem solving with genetic algorithms. They consist of a series of multi-stage instructions inspired by the essential mechanisms of the evolution of living beings and have been shown to be suitable for addressing optimization problems. GA's have some advantages over classical optimization methods, mainly because they can explore many solutions simultaneously, but also drawbacks, due to their relatively lower efficiency. In any case, there are scientific applications solidly based on them and, in addition, they are one of the most beautiful applications of iterative calculation. There are some libraries for Python that facilitate the implementation of the GA, such as Pyvolution, DEAP and Pyevolve.

In the example below, we will solve the same minimization exercise that was presented in section 1.3, following the fundamental steps of the GA (1- Initialization, 2 - selection, 3 - crossover, and 4 - mutation) using directly the basic Python instructions. This example is trivial and simple, as only three genes are involved, but it is appropriate to illustrate this interesting method.

```
#    Genetic Algorithm

import random
from   numpy import *
from scipy import *
from random import randint, uniform
import matplotlib.pyplot as plt

Ncr = 200
Nelite = 190
ngene = 3
ngenerations = 200

#    Fitness function and restrictions
```

```python
def Fu(x1, x2, x3):

    y = (-1)*x1*(x1*x2-sqrt(x3))
    #
    if x1+2*x2-6*x3 > 0: y = -inf
    if x1-x2 >= 8:       y = -inf
    if x1*x2*x3 < 100:   y = -inf
    if x1 < (-10):       y = -inf
    if x1 > 12:          y = -inf
    if x3 < -50:         y = -inf
    if x3 > 50:          y = -inf
    return y

#   Initialize population

f=zeros((Ncr, ngene+1))
n=zeros((Ncr, ngene+1))

for i in range(Ncr):

    x1 = random.rand()*10
    x2 = random.rand()*10
    x3 = random.rand()*10

    f[i,1] = x1
    f[i,2] = x2
    f[i,3] = x3
    f[i,0] = Fu(x1, x2, x3)

#   For each generation ...

fitv = []

for g in range(ngenerations):

    # ... Sort population from low to high fitness

    fsort = f[f[:, 0].argsort()]
    print('g=',g,fsort[-1])
```

```
for i in range(Ncr):

    #   crossover

    #   Discard low fitness chromosomes

    ifather=randint(Nelite, Ncr-1)
    imother=randint(Nelite, Ncr-1)

    # Crossover point is 2
    x1=fsort[ifather, 1]
    x2=fsort[imother, 2]
    x3=fsort[imother, 3]

    # mutate

    x1=x1*(.95+.1*random.rand())
    x2=x2*(.95+.1*random.rand())
    x3=x3*(.95+.1*random.rand())

    # Set Offspring

    n[i,1] = x1
    n[i,2] = x2
    n[i,3] = x3
    n[i,0] = Fu(x1, x2, x3)

  f = n
```

```
g= 0 [-19.933636    2.35482583   4.932130    9.91804844]
g= 1 [-16.724577    2.17250876   4.975966    9.68486987]
g= 2 [-14.142583    2.06611022   4.851424   10.10317408]
g= 3 [-12.920996    2.05540312   4.639764   10.56399667]
g= 4 [-11.595841    1.92571209   4.853851   11.05917124]
...
g= 195 [ 41.260327    8.1460726   0.24598728  49.9691766]
g= 196 [ 41.008998    8.2413695   0.25184911  49.7247078]
g= 197 [ 40.8446103   8.2502391   0.25510765  49.7789149]
g= 198 [ 40.648785    8.1187346   0.2503943   49.5569973]
g= 199 [ 40.636239    8.0970761   0.25021179  49.6266034]
```

The reader can verify that the solution obtained after 200 generations, although not identical, is very similar to the one obtained in section 1.3.

5.6 Neural Networks

In this section we are going to see an interesting machine learning application. Neural networks are inspired by the way information is processed and stored in biological systems. The neurons of living beings receive a multitude of signals through ramifications and, if conditions are sufficient, the neuron activates a signal or nerve impulse driven by the axon. The nervous system, including memory regions, involves the action of a large number of neurons interconnected in a network of enormous complexity.

In contrast, the digital neural networks used in artificial intelligence are comparatively much simpler and organized in layers. In its simplest version, the input signals to a neuron are amplified by weights (w_i), and the output signal or response is increased by a bias (b_i). Normally, at this stage, an activation function is applied to turn an unlimited signal into

"There's been some terrible mistake. I'm programmed for etiquette, not destruction!" (C3-PO during the Battle of Geonosis)

a well-dimensioned or dimensioned signal. The most typical activation function is *sigmoid*, but there are other useful functions such as *Tanh, logistics, RELU, LEAKY*, and others, each with its advantages and disadvantages. This direct process in which input signals

125

are converted into an output signal is known as *Feedforward*.

The awesome feature of a neural network is its ability to learn. When a neural network is initialized, the weights and biases are unknown and therefore the result is an erroneous prediction, a great loss value. But by means of a training process, consisting of a progressive revision of the weights and biases in an algorithm that minimizes the loss known as retropropagation, one arrives at a network capable of making very acceptable predictions. There are several algorithms to minimize losses, such as *stocastic gradient descent (SGD), mini-batch gradient descent, RMSprop*, and others.

In the following example we build a neural network for two input signals with two neurons (x1, x2), a hidden layer with two neurons (h1, h2), and an output signal with only one neuron (o1). The activation signals are the sigmoid of the binary output[0, 1], and the SGD learning algorithm. This neural network is extremely very simple, but adequate to illustrate the four fundamental steps: initialization, learning, feedforward and backpropagation.

The reader is invited to explore the capabilities of Python's powerful automatic learning libraries, such as Kera, PyTorch and Tensorflow.

```python
# Neural Network

from numpy import *

# Data

known = array([[-19,    -9, 1],
               [168,    41, 0],
               [-152, -62, 1],
               [251,   60, 0],
               [-69,  -32, 1],
               [200,   18, 0]])


XX, yy = known[:, :2], known[:, 2]

# Some definitions
```

```
def sigmoid(x): # sigmoid
    y = 1 / (1 + exp(-x))
    return y

def dsig_dx(x): # derivative of sigmoid
    y = sigmoid(x) * (1 - sigmoid(x))
    return y

def L(y1, y2): # loss function
    y = ((y1 - y2) ** 2).mean()
    return y

def ru(): return random.uniform()*2-1

# Neural network processes

eps = []
ls = []


class Neuronet:

  def __init__(W):

    W.w1, W.w2, W.w3 = ru(), ru(), ru()
    W.w4, W.w5, W.w6 = ru(), ru(), ru()
    W.b1, W.b2, W.b3 = ru(), ru(), ru()

  def feedforward(W, x):

    h1 = sigmoid(W.w1 * x[0] + W.w2 * x[1] + W.b1)
    h2 = sigmoid(W.w3 * x[0] + W.w4 * x[1] + W.b2)
    o1 = sigmoid(W.w5 * h1 + W.w6 * h2 + W.b3)
    return o1

  def trainer(W, XX, yy):

    lr = 0.20
    epochs = 1000

    nrecords = len(yy)
    x = array([1,1])
    #
    for epoch in range(epochs):
    #
      for i in range(nrecords):
        #
        x = XX[i]
```

127

```
yactual = yy[i]

# Feedforward

sum_h1 = W.w1 * x[0] + W.w2 * x[1] + W.b1
h1 = sigmoid(sum_h1)

sum_h2 = W.w3 * x[0] + W.w4 * x[1] + W.b2
h2 = sigmoid(sum_h2)

sum_o1 = W.w5 * h1 + W.w6 * h2 + W.b3
o1 = sigmoid(sum_o1)
yypred = o1

# Calculate partials

dL_dypred = -2 * (yactual - yypred)

# Neuron o1
dypred_dw5 = h1 * dsig_dx(sum_o1)
dypred_dw6 = h2 * dsig_dx(sum_o1)
dypred_db3 = dsig_dx(sum_o1)

dypred_dh1 = W.w5 * dsig_dx(sum_o1)
dypred_dh2 = W.w6 * dsig_dx(sum_o1)

# Neuron h1
dh1_dw1 = x[0] * dsig_dx(sum_h1)
dh1_dw2 = x[1] * dsig_dx(sum_h1)
dh1_db1 = dsig_dx(sum_h1)

# Neuron h2
dh2_dw3 = x[0] * dsig_dx(sum_h2)
dh2_dw4 = x[1] * dsig_dx(sum_h2)
dh2_db2 = dsig_dx(sum_h2)

# Recalculate weights

# Neuron h1
W.w1 = W.w1 -lr * dL_dypred * dypred_dh1 * dh1_dw1
W.w2 = W.w2 -lr * dL_dypred * dypred_dh1 * dh1_dw2
W.b1 = W.b1 -lr * dL_dypred * dypred_dh1 * dh1_db1

# Neuron h2
W.w3 = W.w3 -lr * dL_dypred * dypred_dh2 * dh2_dw3
W.w4 = W.w4 -lr * dL_dypred * dypred_dh2 * dh2_dw4
W.b2 = W.b2 -lr * dL_dypred * dypred_dh2 * dh2_db2
```

```
        # Neuron o1
        W.w5 =W.w5-lr * dL_dypred * dypred_dw5
        W.w6 =W.w6-lr * dL_dypred * dypred_dw6
        W.b3 =W.b3-lr * dL_dypred * dypred_db3

        if epoch % 10 == 0:
            yypreds = apply_along_axis(W.feedforward, 1, XX
    )
            loss = L(yy, yypreds)
            print(epoch, loss)
            eps.append(epoch)
            ls.append(loss)
import matplotlib.pyplot as plt
plt.close('all')
plt.plot(eps,ls)
plt.xlabel('Epoch')
plt.ylabel('Loss')

# Train
netw = Neuronet()
netw.trainer(XX, yy)
```

```
0 0.272965704604
10 0.0787910194698
20 0.0462235337578
30 0.0313770360383
...
960 0.000726441279604
970 0.000718548377904
980 0.000710822788982
990 0.000703259285671
```

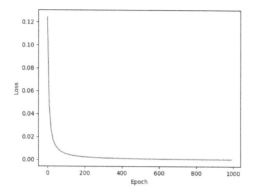

This result is remarkable: based only on the six known data, the system has been able to learn by adjusting over and over again the weights and biases until a very small loss value is reached. Beyond this point, the neuronet barely improves. As an immediate benefit, once the neural network is trained, it can be questioned to make predictions as we see below:

```
# Make some predictions

a = array([-65, -35])
b = array([150, 17])

print(netw.feedforward(a))
print(netw.feedforward(b))
```

```
0.978617158886
0.0304371767509
```

5.7 Big Data

Big Data is a general term referring to the use of a family of analysis techniques applied to data originating from a variety of sources and formats. Data of this nature is generated in large volumes and at ever-increasing speed. This is known as the "three V's". Sometimes the true value of these data may be hidden, and it is necessary to process the information for its usefulness to emerge.

These data sets are so voluminous, even petabytes, and sometimes so unstructured, that conventional data processing software simply cannot handle them. And this trend is increasing more and more, since with the advent of the Internet of Things (IOT), there is an exponential growth in the amount of information of all kinds that must be processed.

Fortunately, recent technological advances have drastically reduced the cost of process computing and data storage to the point of being able to process a large amount of information almost in real time. Both advantages combine to allow a task impossible until now, facilitating decision making, product and service development, predictive maintenance, and exploiting the customer experience. Automated learning is also bringing a hitherto unknown dimension to data analysis techniques.

There is a variety of software to approach Big Data efficiently such as Aqua Data Studio, Microsoft HDInsight, Talend, Splice Machine, among others. With respect to the Python language, a number of libraries such as Pandas, PySpark, SciKit-Learn, Theano, and Keras are useful tools for Big Data analysis. Some of the techniques have already been introduced in the previous sections of this book, but others are specific to handling large volumes of information. However, the type of data and the associated problems are very diverse, so an operational example is beyond the scope of this book.

The reader is invited to consult the specialized literature on Data

Science and Big Data, such as the following magnificent works:

- *Python for Data Analysis*, by Wes McKinney

- *Big Data: Principles and best practices of scalable realtime data systems*, by Nathan Warz

- *Data Science for Beginners*, by Leonard Deep

Computer scientist Katie Bouman reacting when the historical image of the Powehi black hole was generated with the algorithm invented by her team, based on artificial intelligence techniques. K. Bowman.

This young man is the Turkana's boy, who with his 850 cc brain, sprinted on the evolution race 1.6 million years ago to pass the torch of intelligence to later hominids and finally to us, his successors.

Credits: Musée National de Préhistoire, in Les Eyzies-de-Tayac-Sireuil.   Photo by W. Sauber, licensed under the Creative Commons Attribution-Share Alike 4.0 International licensed under the Creative Commons Attribution-Share Alike 4.0 International

Chapter 6

Control Methods

In this section we will explore
some of the possibilities of Python in the slippery sand of
the theory of stability and control of dynamic systems. These
systems are presented in a wide
variety of fields of physics and
engineering, from mechanical,

electrical, hydraulic, chemical or even nuclear, to name a few, but
also biological, social or economic systems. It is common to ask
whether a given system is stable, what individual characteristics it
must have to provide a desired dynamic response, either as a single
system or combined with elements of compensation or control.

The study of the stability of a dynamic system can be approached
according to its behaviour in the time domain or in the frequency
domain. In this section we will see a combination of both. Some
of the material comes from the excellent tutorial prepared by the
University of Michigan and Carnegy Melon:

http://ctms.engin.umich.edu/CTMS/index.php?aux=Home.

6.1 Methods in the Frequency Domain

6.1.1 Welcome aboard!

1. Is this plane stable?

To illustrate some of the methods of classical control theory, let's focus on an interesting case. Suppose we want to design a control system to define the angle of attack of a passenger plane. The pilot can act on the elevation flaps at the rear and act in combination with the horizontal stabilizers. Thus, a command signal δ induces a dynamic behavior in the plane and an angle of attack θ of the plane, through a transfer function given by:

$$\frac{\Delta\theta}{\Delta\delta} = \frac{1.151s + 0.1774}{s^3 + 0.739s^2 + 0.921s}$$

Input δ and output Θ

In the following Python script we introduce the transfer function in open loop, analyze the poles and zeros of the dynamic system, and get the response to a step and an impulse of the input.

```
# Jet Airliner

from control import *
import matplotlib.pyplot as plt
from numpy import *

#   Open loop Transfer Function

P = tf([1.151, 0.1774],
       [1, 0.739, 0.921, 0])

print("P = ", P)

#   Poles and zeros

fig = plt.figure(1)
P_poles, P_zeros = pzmap(P)
print("P_Poles ", P_poles)
plt.title('Poles and Zeros of P(s)_OL')

#   step response of 0.2 rad

fig = plt.figure(2)
time = np.arange(0, 40, 0.1)
t, P_OL_step = step_response(0.2*P, time)
plt.plot(t, P_OL_step)
plt.ylabel('P_OL_step')
plt.title('Step response of P(s)_OL')

#   impulse response

fig = plt.figure(3)
t, P_OL_impulse = impulse_response(0.2*P, time)
plt.plot(t, P_OL_impulse)
plt.ylabel('P_OL_impulse')
plt.title('Impulse response of P(s)_OL')
```

```
P =
1.151  s  +  0.1774
-------------------------
s^3  +  0.739  s^2  +  0.921  s

P_Poles   [-0.3695+0.8857j  -0.3695-0.8857j   0.0000+0.j]
```

The poles of the system, i.e. zeros of the denominator, are either zero or real-negative indicating that the system is marginally stable. To understand how the dynamic response to an input signal is, it is useful to inspect the step response and impulse response. With the first one we observe that the command of a rear pitch angle puts the plane at an increasingly higher angle. In the case of commanding an instantaneous impulse, the plane first draws an important initial oscillation and then ends up stabilizing at a non-zero elevation angle. This means that there is a wrong angle remaining.

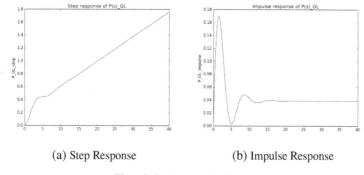

(a) Step Response   (b) Impulse Response

These behaviors are inadequate.

To avoid this undesirable behavior, we are going to modify the system by closing the loop and installing a compensator or a control.

2. Gain Determination - Root Locus

By closing the loop and feeding back the output, you can control the error effectively. In its simplest version, the controller consists only of a proportional multiplier of the error signal, $C(s) = K_p$. Whether the $H(s)$ feedback is a negative-unit or a more complex function, the closed-loop transfer function can always be obtained such as

$$Y(s)_{cl} = \frac{P(s)}{1 + K_p P(s) H(s)}.$$

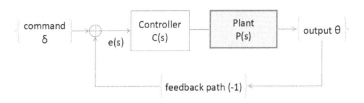

Closing the loop to achieve controlability

Let us look at the denominator: $1 + K_p P(s) H(s) = 0$. It is the characteristic equation of the system and the sign and magnitude of its roots determine the dynamic behavior. The *Root Locus* method is a graphical method that draws the location of the roots of this equation in the complex plane sweeping K_p from 0 to infinity and thus determining the range of values of K_p that make the closed-loop system stable.

In the attached script, we first get the root position of $1 + K_p P(s) H(s) = 0$ finding that, in the example, all roots always remain on the negative side which indicates that the closed-loop system is stable for any value of K. Next, the script calculates the response to a step of the input signal showing that, although the response is stable, the higher the value of the K_p multiplier, the more oscillating the response, to the point that the control specification cannot be met. In the next subsection we are going to look for another type of compensation.

```
#... continuation

# Root Locus of equation 1+k*P(s) = 0

fig = plt.figure(4)
[rlist, klist] = rlocus(P, grid=True)

#   Closed Loop with various gains

K = 0.3 ; P1 = K*P/(1+K*P)
K = 1   ; P2 = K*P/(1+K*P)
K = 10  ; P3 = K*P/(1+K*P)

#   step response of 0.2 rad

fig = plt.figure(5)

t, P1_step = step_response(0.2*P1, time)
t, P2_step = step_response(0.2*P2, time)
t, P3_step = step_response(0.2*P3, time)


plt.plot(t,P1_step, t, P2_step, t, P3_step)
plt.ylabel('CL Step response with several gains')
plt.legend(['K=0.3', 'K=1', 'K=10'])
```

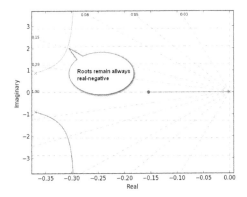

No root is positive, indicating that the system is stable for any K value.

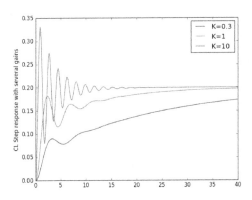

The higher the K value, the faster the feedback system responds, but the trend to oscillate also increases.

3. PID Controller

This type of controllers is currently very widespread in the industry due to its ease of implementation and generally satisfactory behavior.

The PID control has three terms: the *proportional* term (Kp), which gives sensitivity to the controller, the *integral* term (Ki/s), which accelerates the movement and eliminates the error in steady state although it can cause some overshoot, and the *derivative* term ($K_d s$), which stabilizes and flattens the response. Also, for signals mixed with noise, the latter term can be complemented with a high frequency filter τ_d.

$$C(s) = K_p + K_i/s + \frac{K_d s}{\tau_d s + 1}$$

There are several techniques of varying complexity to determine the values of Kp, Ki, and Kd, such as manual adjustment, the heuristic method of Ziegler-Nichols, the Cohen-Coon method, the Astrom-Hagglund method, and others.

In the exercise that we are following, we will consider that we have the following specification for the controlled system for a step reference of 0.2 radians:

- Overshoot less than 10%

- Rise time less than 2 seconds

- Settling time less than 10 seconds

- Steady-state error less than 2

In our example, we have determined the three constants manually, by pure trial and error until the step specification is met. The PID controller is as follows:

$$C(s) = 5 + \frac{1.7}{s} + 3s$$

```
#... continuation

kp = 5
ki = 1.7
kd = 3
taud = 0

# Define OL transfer function with PID

PID = tf([kd + kp*taud, kp+ki*taud, ki],
         [taud,1, 0])

Ppid_CL = PID*P/(1+PID*P)

#   step response with PID controller

fig = plt.figure(6)
t, Ppid_CL_step = step_response(0.2*Ppid_CL, time)
plt.plot(t, Ppid_CL_step)
plt.ylabel('Ppid_CL_step')
```

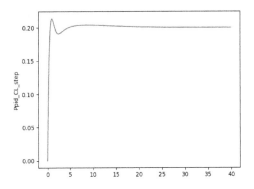

The plane with a PID controller already meets the specification.

4. Lead/Lag Compensator

Another way to achieve an adequate response is to introduce a lead/lag compensator. The lead value is useful to improve the stability and speed of response, and the lag is to eliminate the error in steady state. Their values are usually determined by techniques of root locus or with the frequency response method. In the case of our example, we arrive at the following compensator:

$$C(s) = K\frac{Ts + 1}{\alpha Ts + 1}$$

With the attached script we calculate the response to a step and check that is a quick response and that meets the specification.

```python
#... continuation

K = 10
T = 0.55
alpha = 0.04

# Define OL transfer function

Compensator = K*tf([T,1],
                    [alpha*T,1])

Pcomp_CL = Compensator*P/(1 + Compensator*P)

#   step response with lag compensator

fig = plt.figure(7)
t, Pcomp_CL_step = step_response(0.2*Pcomp_CL, time)
plt.plot(t,Pcomp_CL_step)
plt.ylabel('Pcomp_CL_step')

#   All together

fig = plt.figure(8)
plt.plot(t,Pcomp_CL_step,'--', t, Ppid_CL_step,
         '-', t, P2_step, ".")
```

```
plt.legend(['with compensator', 'with PID',
'CL alone'], loc="lower right")
```

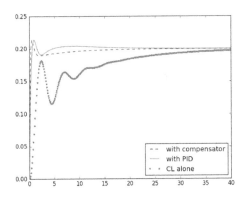

Comparison of the three methods: closed loop, PID, and lead/lag.

The control module includes also many useful functions that help in analyzing and stabilizing dynamic systems, like Bode, Nichols, and Nyquist, among others.

Proposed exercises

- **Stability Gain Range.** Assume a system with the following open loop transfer equation:

$$G(s) = \frac{s+1}{s(s+2)(s+4)^2} = \frac{s+1}{s^4 + 10s^3 + 32s^2 + 32s}$$

Close the loop with a negative unit feedback and show that the controlled system is stable in the range K=1.9 to k=200. *Suggestion:* plot the root locus of the characteristic equation

145

1+kG=0, and right click in the vicinity of the roots curve to obtain the desired information.

- **DC - Motor Speed**

Assume that a motor controlled by electric field has the following transfer function

$$P(s) = \frac{\dot{\Theta}(s)}{V(s)} = \frac{1000}{50s^2 + 600s + 1001}$$

This exercise consists of designing a PID controller to ensure a given angular velocity in a reference. For practical reasons, the desired operating specifications for a step input of 1 rad/sec are the following:

- Settling time less than 2 sec,
- Overshoot less than 5%, and
- steady state error less than 1%.

Test different combinations of values. Try also (Kp=75, Ki=1, Kd=1) and (Kp=100, Ki=200, Kd=10).

- **PID tunning with Ziegler-Nichols.** A popular heuristic method for tunning a PID controller is the one due to John Ziegler and Nathaniel Nichols. Asume we have a system with the following plant transfer function:

$$P(s) = \frac{1}{(s+1)^3} = \frac{1}{s^3 + 3s^2 + 3s + 1}$$

1. Set Ki=Kd=0, and increase Kp until the loop starts to oscillate (Ku). Take note of the period Pc.
2. Now set Kp=0.5 Ku, $Kd = KpPc/8$, and $Ki = 2Kp/Pc$.

6.2 Methods in the Time Domain - The State Space

Although the frequency domain method is extremely useful and is one of the main tools in control engineering, its applicability is limited when there is more than one input and output signal.

In contrast, time-domain techniques can easily be used for non-linear systems that vary in time and involve multiple variables. Their representation is a fundamental basis for modern control theory. A dynamic system can be described as

$$\dot{\mathbf{x}}(\mathbf{t}) = A \cdot \mathbf{x}(\mathbf{t}) + B \cdot \mathbf{u}$$
$$\mathbf{y}(\mathbf{t}) = C \cdot \mathbf{x}(\mathbf{t}) + D \cdot \mathbf{u}$$

where $\mathbf{x}(\mathbf{t})$ is the *state* vector, $\mathbf{y}(\mathbf{t})$ is the output vector, $\mathbf{u}$ is the input or *control* vector, and A,B,C,D are matrices. To control a state-space system, a reference $r(t)$ is set, and the state variables are fed back with the multiplier matrix $\mathbf{u_f} = -K_c \cdot \mathbf{x}(\mathbf{t})$. In this section we present a method to determine the K_c multiplier and the eventual N precompensator with Python.

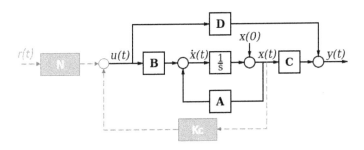

State vector $x(t)$ with feedback and precompensator.

Dynamics of a Motor

1. Is the system stable?

We are going to assume that we have a cd engine controlled by armor, whose dynamics are governed by:

$$\frac{d}{dt}\begin{bmatrix} \dot{\theta} \\ i \end{bmatrix} = \begin{bmatrix} -10 & 1 \\ -0.02 & -2 \end{bmatrix}\begin{bmatrix} \dot{\theta} \\ i \end{bmatrix} + \begin{bmatrix} 0 \\ 2 \end{bmatrix} V$$

$$y = \begin{bmatrix} 1 & 0 \end{bmatrix}\begin{bmatrix} \dot{\theta} \\ i \end{bmatrix}$$

where i is the armature current, $\dot{\theta}$ is the shaft angular velocity (output), and V is the voltage source (input). Besides, let's suppose we want such a controller that for a 1-rad/sec step reference input, the design criteria are the following:

- Settling time less than 2 seconds

- Overshoot less than 5%

- Steady-stage error less than 1%s

In the following lines analyze the system and determine the controller that meets the specifications. This system is analyzed in the following script:

```
# Control motor-speed

from numpy import *
from control import *
import matplotlib.pyplot as plt

global A,Q

# Setup the system
```

```
A = array([[-10.   ,    1.  ],
           [ -0.02,   -2.  ]])
B = array([[ 0.],[ 2.]])
C = array([1,0])
D = 0
motor_ss = ss(A, B, C, D)

# Stability, Observability and controlability

print("Eigvals ", linalg.eigvals(A))

observability   = obsv(A, C)
controlability1 = ctrb(A, B)

print('Rank is ', linalg.matrix_rank(
        controlability1))
print('Det of controlability matrix is ', \
      linalg.det(controlability1))
```

```
Eigvals  [-9.99749922 -2.00250078]
Rank is  2
Det of controlability matrix is   -4.0
```

Since the eigenvalues of the system matrix A are all real-negative, we know that our system is stable. Besides, since the controllability matrix is full rank (det is not zero) the system is controllable. Now, we will determine Kc

2. Regulator K_c

```
# placing the poles of closed loop controller

p1 = -5 + 1j;
p2 = -5 - 1j;
```

149

```
Kc = place(A, B, [p1, p2])

print('Kc ',Kc)
```

```
Kc  [[12.99 -1.  ]]
```

```
sys_cl = ss(A-B*Kc, B, C, D)

# Plot step response

T, sys_cl_step = step_response(sys_cl,
                               arange(0, 3, 0.01),0)
plt.figure(10)
plt.plot(T, sys_cl_step)
plt.title('motor_ss closed loop')

print("Last step value is ",sys_cl_step[-1])
```

```
Last step value is  0.0769230826985
```

3. Precompensator $\overline{N}$

We observe that the closed-loop system meets the specification, but does not reach the target reference value 1 rad/s, so we add a pre-compensator N of $1/0.0769 = 13$. By doing this, the step response already meets all the requirements of the specification.

```
Nbar = 13
#
# Step response

T, y_sys_cl_N = step_response(sys_cl*Nbar,
```

```
                          arange(0, 3, 0.01),0)


# Plot step response

T, yout = step_response(sys_cl*Nbar,
                         arange(0, 3, 0.01), 0)
plt.figure(1); plt.plot(T,yout)
plt.title('motor_ss closed loop')
```

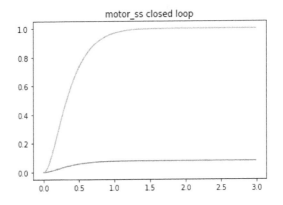

Step response with regulator (blue) and also a precompensator (orange)

6.2.1 Dynamics of an Airplane

1. Is the system stable? Controllable?

Now we are going to analyze the dynamics of the Jet airliner from example 6.1.1. The differential equations of motion are as follows:

$$\frac{d}{dt}\begin{bmatrix} \alpha \\ q \\ \theta \end{bmatrix} = \begin{bmatrix} -0.313 & 56.7 & 0 \\ -0.0139 & -0.426 & 0 \\ 0 & 56.7 & 0 \end{bmatrix} \begin{bmatrix} \alpha \\ q \\ \theta \end{bmatrix} + \begin{bmatrix} 0.232 \\ 0.0203 \\ 0 \end{bmatrix} \begin{bmatrix} \delta \end{bmatrix}$$

$$y = \begin{bmatrix} 0 & 0 & 1 \end{bmatrix} \begin{bmatrix} \alpha \\ q \\ \theta \end{bmatrix}$$

```
# Jet Airliner

from numpy import *
from control import *
from numpy.linalg import matrix_rank


import matplotlib.pyplot as plt

# Define the open loop state space system

A = matrix([[-0.313,   56.7,      0],
            [-0.0139,  -0.426,    0],
            [0,        56.7,      0]])

B = matrix([[0.232], [0.0203], [0]])

C = matrix([0, 0, 1])

D = matrix([[ 0 ]])

SSplane_ol = ss(A, B, C, D)

# let us look at the eigenvalues
print("Eigenvals of A ", linalg.eigvals(A))

# Plot OL step response
T, YSS_ol_step = step_response(0.2*SSplane_ol,
                               time, 0)
plt.figure(20)
```

```
plt.plot(T, YSS_ol_step)
plt.title('YSS_OL_step')

# Observability and controlability
observability = obsv(A,C)
controlability1 = ctrb(A,B)
print('Controlability matrix is ', controlability1)
print('Rank is ', linalg.matrix_rank(
        controlability1))
print('Det of contr matrix is ', linalg.det(
        controlability1))
```

```
Eigenvals of A
  [ 0.0000+0.j -0.3695+0.88596713j -0.3695-0.88596713j]
Controlability matrix is
   [[ 0.232       1.078394    -1.01071374]
   [ 0.0203      -0.0118726   -0.00993195]
   [ 0.          1.15101      -0.67317642]]
Rank is  3
Det of contr matrix is  -0.00437266444181
```

Since the controllability matrix is 3x3, the rank has to be 3. Besides, we know that our system is controllable since the controllability matrix is full rank (the determinant is not zero).

2. Regulator K_c and Precompensator $\overline{N}$

```
# Calculate the state space optimized [K] and [N]

from numpy import *
from scipy import *
from scipy import integrate
from scipy.optimize import minimize


global over
```

```python
Yref = 0.2

#  Define IAE - Integral of the absolute error (IAE)


def IAE_step(p):
    global over
    k1, k2, k3,nbar = p
    K = array([k1, k2, k3])
    sys_cl = ss(A-B*K, B, C, D)
    T, YSS_cl_step = step_response(Yref*sys_cl*nbar,
                                   time,0)
    er = Yref - YSS_cl_step
    IAE = integrate.cumtrapz(abs(er), time,
                             initial = 0)
    return IAE[-1]


#  Define constraint for overshoot < 1.05


def f1(p):
    global over
    k1, k2, k3,nbar = p
    K = array([k1, k2, k3])
    sys_cl = ss(A-B*K, B, C, D)
    #
    T, YSS_cl_step = step_response(Yref*sys_cl*nbar,
                                   time,0)
    over = max(YSS_cl_step/Yref)
    return -over+1.05

#  Define constraint for settling time < 10 s

def f2(p):
    k1, k2, k3, nbar = p
    K = array([k1, k2, k3])
    sys_cl = ss(A-B*K, B, C, D)
    T, YSS_cl_step = step_response(Yref*sys_cl*nbar,
                                   time,0)
    #
    slope10 = (YSS_cl_step[101]\
               -YSS_cl_step[100])/0.1
```

```
    return -abs(slope10)+0.0001

#  Minimize the IAE with restricted overshoot

constraints = [{'type':'ineq', 'fun':f1},
               {'type':'ineq','fun':f2}]
results = minimize(IAE_step, (1, 1, 1, 1),
options = {'maxiter':200},
constraints=constraints)

# Print results for optimized system

k1, k2, k3, nbar = results.x
K = array([k1, k2, k3])
print('[K]=',K, 'Nbar=', nbar)

sys_cl = ss(A-B*K, B, C, D)
T, YSS_cl_step = step_response(Yref*sys_cl*nbar,
                               time,0)
plt.figure(14)
plt.plot(T, YSS_cl_step)
plt.title('YSS_CL_step for IAE min')
```

```
[K]= [ -5.07465031e+00    7.91544473e+03    4.99106358e+02]
Nbar= 499.620642432
```

6.3   Lyapunov Stability

In the previous sections we have studied the stability of dynamic systems with methods of frequency domain and time domain. These methods, especially the latter, are widely used because they are easy to apply and because they are quite intuitive. In this section we deal with Lyapunov's concept of stability, which approaches the problem with an ingenious approach because it allows us to study the

155

stability of a system of differential equations and estimate the different regions in the phase plane, *without the need to solve the system in time*. Although on the one hand, it is a less direct method, on the other hand it allows the formal study of dynamic systems based solely on their mathematical properties.

In a simplified way, we will say that an equilibrium point X_0 of the homogeneous differential equation $\dot{X} = f(X)$ is stable if, after a disturbance of the initial condition, all the solutions to the equation that begin in the vicinity of this point X_0 remain close to it forever. This definition of stability is named after Aleksandr Liapunov, who published his doctoral thesis *The General Problem of Stability of Motion* in 1892.

In his method, the challenge is to find a positive function arbitrarily defined as $V(X)$, which meets the following requirements

$$\dot{V} = \frac{V(X))}{dt} = \nabla V \cdot f(X) \leq 0.$$

The existence of this function implies that the system $f(X)$ is asymptotically stable in the sense of Lyapunov, and the region Ω where this is fulfilled, is the basin of stability. In the end, this $V(X)$ function is related to the concept of potential energy of a system.

Lyapunov studies stability with two methods. The first one consists of analyzing the signs of the eigenvalues of the Jacobian of $f(X)$, associating stability to the non-existence of real-positive eigenvalues. The second method gravitates around the definition given in the previous paragraph. For practical purposes, for linear systems, it is a mater of solving the following equation:

$$A^T P + P A = -Q$$

where A is the Jacobian of $f(X)$, Q is an arbitrary positive defined matrix, an P the unknown matrix to be determined. Finally, $V(x) = X^* P X$.

In the following exercise, we study the stability of a system whose Jacobian is given by

$$A = \begin{bmatrix} -1 & 3 \\ 0 & -1 \end{bmatrix}$$

```
#   Lyapunov stability

#   System matrix

A = array([[-1, 3],
           [0, -1]])

# Trial matrix

Q = array([[1, 0],
           [0, 1]])

print("eigvals of A ",linalg.eigvals(A))
```

```
eigvals of A   [-1. -1.]
```

According to the first method of Lyapunov, since all eigenvalues of the Jacobian are real-negative, the system is asymptotically stable. Now, let us investigate whether a region of stability exists.

```
# ... continuation.

from scipy.optimize import fsolve

# find the P matrix in Lyapunov eqn

def lyapunov4(p):

    p11, p12, p21, p22 = p
    P = matrix([[p11, p12], [p21, p22]])
```

```
    #
    em = A*P + P*A.transpose() + Q
    #
    return em[0,0], em[0,1], em[1,0], em[1,1]

# Obtain P matrix

p11, p12, p21, p22 = fsolve(lyapunov4,(1,1,1,1))
P = matrix([[p11, p12], [p21, p22]])
print("eigvals of P ", linalg.eigvals(P))
```

```
eigvals of P   [ 2.97708173   0.27291827 ]
```

Both eigenvalues are positive, indicating that function $V(X)$ is positive defined, and its time derivative is negative. Therefore, the region is a stability basin. Let's see what this function looks like:

```
# ... continuation

# Plot 3D Surface

from mpl_toolkits.mplot3d import Axes3D
from matplotlib import cm

fig = plt.figure(12)
ax = fig.gca(projection='3d')

# Make grid

X1 = np.arange(-5, 5, 0.5)
X2 = np.arange(-10, 10, 0.5)
X1, X2 = meshgrid(X1, X2)
Z = p11*X1**2 + X1*X2*(p12+p21)+p22*X2**2

# Plot surface

ax. plot_surface (X1 , X2 , Z ,
```

```
                    cmap =plt.cm.jet ,rstride =1,
                    cstride =1, linewidth =0)
plt. xlabel ("x1")
plt. ylabel ("x2")
plt.show()

# Plot contour map

from matplotlib.pyplot import contourf, colorbar
fig = plt. figure (18)
contourf (X1 ,X2 ,Z)
colorbar()
plt. xlabel ("x1")
plt. ylabel ("x2")
plt.show()
```

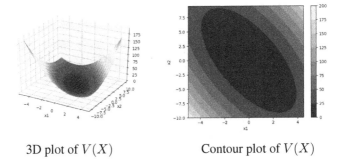

3D plot of $V(X)$ Contour plot of $V(X)$

Think of all the beauty still left around you and be happy [...] I don't want to live in vain like most people. I want to be useful or bring enjoyment to all people, even those I've never met. I want to go on living even after my death! (Anne Frank, 1944)

Appendices

Appendix A

Python Primer

This part is dedicated to those readers who have little or no knowledge of Python as a programming language. Certainly, people with a real interest in something build knowledge from concrete information, that is, they apply an inductive method with easy-to-understand examples that synthesize some teaching. In many occasions, it is enough to introduce the key information or to give some clues about where to start walking freely.

In the following lines, the most elementary of Python will be presented: how to execute the code, perform basic calculations, write and execute a script, and things like that. Of course, if the reader already knows this language, he or she can skip this appendix without hesitation.

The basics

Python is a programming language for high-level scripts with very interesting features: simplicity, power and versatility. Guido van Rossum, who created the language, insisted that the language must favour the writing of legible and well-structured scripts.

It is a powerful and fashionable language, such as Java or C, widely used in academic and business environments, but much more intuitive and simple. It can be used for web development, GUI development, scientific and numerical calculation, software development and operating system administration.

Python is multiplatform: It can run on Windows, Linux, and macOS. Commands and statements can be executed interactively in console mode, or they can be stored and executed in a *.py script file. Console mode is very useful for testing instructions before writing them into the script.

There is a lot of information (tutorials, manuals, user groups, Internet examples, etc.). The code, utilities, manuals, libraries, and everything related to Python are in www.python.org. Also, you will find a very good description in
https://es.wikipedia.org/wiki/Python..

Running Python

Today, there is a number of suites that greatly facilitate writing and running Python. For MS-Windows, the Anaconda/Spyder IDE, which is an integrated development environment for the Python language, i highly recommended. The suite is developed and distributed under the MIT license, and it is multi-platform and free.

Anaconda/Spyder integrates useful libraries (modules) such as NumPy, SciPy, and Matplotlib, and it gives easy access to many others like Tkinter, Pandas,...

Basically, it shows three windows: (1) edition, (2) variables and files, and (3) execution console. Do you want to run your first script?

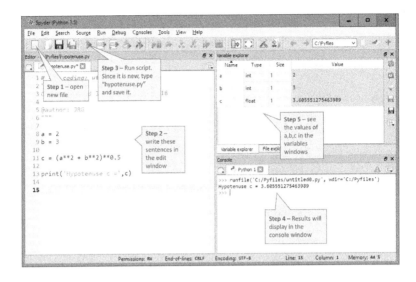

Your first script!

Assignment of variables

Variables can be any combination of characters and numbers, uppercase, and lower case. Assigned values can be reals or floats, integers, complex, character chains or strings, etc.

```
# This line is a comment

" " "
This block is a LONG COMMENT. It is
bounded between triple quote marks.
As many lines as you want.
" " "


mass = 36.12       # a real number (float)

gravity = 9.81     # another real number

jas = 32           # an integer
```

165

```
alpha = 2 + 3j    # a complex number

beta = complex(2,3) # alternate definition of
    complex

My_name = "Wonder123_4dogs" # alphanumeric string
```

Assignment of lists, dictionaries, and tuples

Lists of numbers and alphanumeric strings can be created easily.

```
# Lists

primes = [2, 3, 5, 7, 11, 13]

days = ['monday', 'tuesday', 'Friday']

Movies =    [["Star Wars", 1977],
             ["Superman II", 1980],
             ["Gremlins", 1984],
             ["Schindler's List", 1993]]


# Dictionaries

my_dictionary = {
    'cat': 'Frisky',
    'dog': 'Spot',
    'fish': 'bubbles'
}

# Tuples - are lists with fixed content
#          They cannot be modified.

tup1 = ('physics', 'chemistry', 1997, 2000)

tup2 = (1, 2, 3, 4, 5)
```

Arrays are like vector and matrix structures, and they can be operated in algebraic operations as numbers. Before using arrays, you need to import numpy library.

```python
# Arrays

from numpy import array

vec = array([1,4,6])

# Column array
vec_col = array([1,2,3]).reshape(-1,1)

# Column array, other method
vec_col1 = array([[1],[2],[3]])

 # Column array, more visual
vec_col2 =  array([[1],
                   [2],
                   [3]])

# Matrix
mat = array([[1,2,6], [3,4,9],[1,-2,7]])

# Matrix, more visual
mat1 = array([[1,2,6],
              [3,4,9],
              [1,-2,7]])
```

Calculations and operations

You can perform all sort of mathematical operations with data. They can be reals, integers, text chains, etc.

```python
from scipy import *
from numpy import *
```

```
force = mass * gravity

alpha = log((sin(pi*gravity))**2)-0.5

Ux3 = pi*sin(alpha)*e

Vec2 = vec * vec

Vec3 = vec2 * 2/3

Mat3 = mat1 * mat1

Au = linalg.eigvals(mat1)

your_name = "hello "+ My_name + "abc"
```

Extract data from arrays

It is a very powerful syntax. Note that array indexing starts in 0, not in 1.

```
aa = vec[2]      # extract 3d coordinate of array

bb = vec[0:2]    # extract from 1st to 3rd coordinate

cc = vec[:]      # extract all

Evec = mat1[:, 1]  # extract 2nd column

last = vec[-1]   # get the last in array
```

Functions

A very useful feature in Python is the possibility of defining functions. This is a good strategy when you need to execute a series of sentences that can be arranged in a callable structure. Besides, they add clearness to the reading. Functions can have any number of inputs and outputs as can be seen in the following examples.

```
# Example with one input, one output

def conductivity(t):
    a = 2150
    b =  1.05
    y = a + b/((t+273)-73.15)
    return y
```

```
# Example with two inputs, three outputs

def hipot(x,y):
    z = (x**2 + y**2)**0.5
    l = x + y + z
    v = x*y*z*l
    return z,l,v
```

Once they have been declared, they can be called downstream in the script, in the same way as any other intrinsic function.

```
...
T = conductivity(2200) / 235.12

gamma = 6.12*hipot(2,3)
...
```

Everything that has been declared inside a function is local. To become accessible you need to include it in the return sentence, or declare it as global. However, any variable that has been declared out of the function bounds is accessible from the inside.

Importing modules (libraries)

There are many libraries in Python that allow you to perform calculations in different areas. Some of the most useful are scipy, numpy, and matplotlib, and they are included in the Anaconda package, so you only need to import the library from the script.

In order to see which modules are installed, open a command prompt and type "> pip list". If you need a module that is not installed, open a command prompt and type "> pip modulename". Check *https://docs.python.org/3/py-modindex.html* for a list of available modules.

You can import an entire module or only a portion of it. You will find different practices on the Internet. We can simplify up to four modes of importing a molule:

```python
# 1 - Not bad, it imports the entire module
from numpy import *

# 2 - Import only a specific function
from numpy import array

# 3 - Import the entire module
import numpy

# 4 - Import the module and rename it
import numpy as np   # 4 - import and rename it
```

Calling a portion depends on how we imported the module

```python
g = array([1,2])         # First and second method
g = numpy.array([1,2])   # Third method
g = np.array ([1,2])     # Fourth method
```

Loops

Scripts often need some internal flow controls. In this section we see examples of while, for, and if/then/else loops. Note that all sentences inside a loop are indented, four spaces per looping level

```
#   While loop

i = 0
while i < 5:
    print(i)   # print numbers from 0 to 4
    i = i+1    # i +=1 is an alternative


# For loop

for i in (0, 1, 2, 3, 4):    # or in range(4)
    print (i)  # print numbers from 0 to 4


#  If / then / else loop

a = 1
b = 2

if a == b:
    print('They are identical')
else:
    print ("They are different")
```

Plotting 2D arrays

```
#   Data

x1 = array([2.1, 3.2, 6, 9.5, 10, 12])
y1 = array([4.1, 6., 37, 70, 92, 100])
```

```
#    The graph

import matplotlib.pyplot as plt # plotting module

plt.close('all') # Let us close all previous plots
fig = plt.figure(1) # number the figure
plt.plot(x1,y1,'g-o', x1, y1*2,'r--x')

plt.title('Plot of x1 versus y1')
plt.legend(('y1','2*y1'))
plt.xlabel('x1, meters')
plt.ylabel('y1, pounds')
```

Note that the type of connections between data points is controlled with 'g-o' (green/line/circle) and 'r–x' (red/dash/x). There are many combinations of colors, lines, and marks. See Python manual for options.

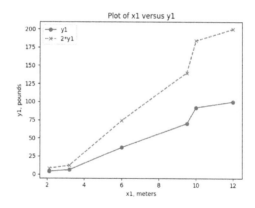

Plotting 3D arrays

```python
from numpy import array

# Data

x1 = array([2.1, 3.2, 6, 9.5, 10, 12])
y1 = array([4.1, 6., 37, 70, 92, 100])
z1 = array([6., 6., 9., 15., 14., 13.])

# Now, we plot the 3D line

from mpl_toolkits.mplot3d import Axes3D
import matplotlib.pyplot as plt

fig = plt.figure(2)
ax = plt.axes(projection='3d')
ax.plot(x1, y1, z1, '-o')

plt.title('Example of 3D line plot')
plt.xlabel('x1 values')
plt.ylabel('y1 values')
```

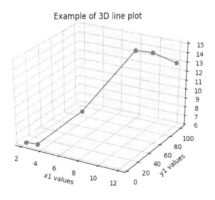

Example of 3D line plot

173

Plotting a Contour plot

```python
# Let us prepare data

from numpy import arange, meshgrid
from scipy import exp

# Generate high resolution x and y axis and grid
x= arange(-2,2,.05)
y= arange(-2,2,.05)
xx,yy = meshgrid(x,y)

# Assign zz for each mesh point
zz = xx*exp(-xx**2 - yy**2)

# Now you can plot
from matplotlib.pyplot import contourf, colorbar
fig = plt.figure(3)
contourf(xx,yy,zz)
colorbar()
plt.title('Example of 3D contour plot')
plt.xlabel('xx values')
plt.ylabel('yy values')
```

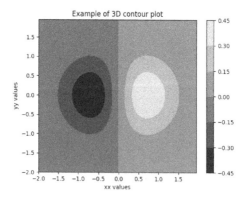

Plotting a 3D surface

```python
import matplotlib.pyplot as plt
from matplotlib.pyplot import axes, contourf,
    colorbar
from mpl_toolkits.mplot3d import Axes3D

fig = plt.figure(4)
ax = plt.axes(projection='3d')
ax.plot_surface(xx, yy, zz, cmap=plt.cm.jet, rstride
    =1, cstride=1, linewidth=0)

plt.title('Example of 3D surface plot')
plt.xlabel('xx values')
plt.ylabel('yy values')
```

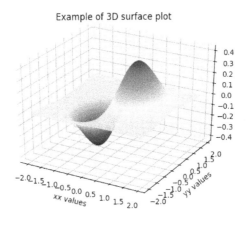

Reading data from *txt* files

First, we read the entire file and store it in a list

```
# Read the txt file

arch1 = open("Humanrights.txt","r")

Lines = arch1.readlines()

arch1.close
```

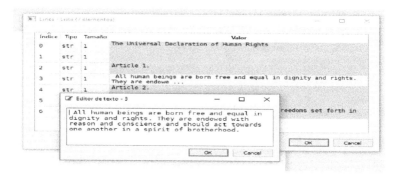

Reading from a text file

Once the file has been read and stored in 'lines' we can search inside, extract values, and so on. Some interesting things to do are:

```
# Count the number of lines
Lon = len(lines)

# Read text in line 3, columns 0 to 38
A = lines[3][0:38]

# Read float number in line 3, columns 3 to 15
AA = float(lines[12][3:15])
```

```
# Find line with with a specific text
i=0 ; click=0
for i in range(Lon-1):
    if click==0 and ("right to live" in lines[i]):
    ilin=i ; click=1

# Extract lists of values X and Y
X,Y = [],[]
for i in range(12,19): # between lines 12 and 19
    X.append( float( lines[i][1:5] ))  # columns 1
    to 5
    Y.append( float( lines[i][6:12] )) # columns 6
    to 12
```

In the previous example, you need to know in advance the line numbers that include your data. But, if you already know that numbers are in fields 0 and 1, the last two lines are simplified:

```
X.append( float( lines[i].split()[0]))
Y.append( float( lines[i].split()[1]))
```

Read data from *txt* file between some headers

Occasionally, your data is inside a file between two headers. If you were to grab the data manually, first you should search for the begin header, descend to where data start actually, then copy all the data you need, and paste in a list or array. With Python this can be automatized greatly. You have to tell the code the same orders as you would do manually.

```
#    1- Open, count, and read lines in file
myfile = 'c:\pyfiles\filedata.txt'
num_lines = sum(1 for line in open(myfile))
with open(myfile) as f: lines = f.readlines()
```

177

```
#      2- Find lines with begin and end headers
click=0
for i in range(num_lines-1):
    if click==0 and ("begin text" in lines[i]):
        istring1 = i; click=1
    if click==1 and ("end text" in lines[i]):
        istring2 = i ; click=2


#      3- Extract information in selected fields.
#         Sometimes you need to adjust lines
TIMEv=[]
VPWIv=[]
for linea in range(istring1+3 , istring2-2):

    time= float(lines[linea].split()[0])
    vol = float(lines[linea].split()[5])

    TIMEv.append(time)
    VPWIv.append(vol)
```

Read data from *txt* file with pure columns

If data is arranged in pure columns in a text file, the reading is very simple. In the example below, data are stored in several columns. We want to read data in columns 1,2, and 5.

```
arch2 = open("columns.txt", "r")
t,x,y = [],[],[]

for l in arch2:
    row = l.split()
    print(row)
    t.append(float(row[0]))
    x.append(float(row[1]))
    y.append(float(row[5]))
arch2.close()
```

178

Read data from an Excel *xls* file

Something that is extremely useful when performing analysis is the possibility to read data from an Ms-Excel spreadsheet.

For cleanliness, in the example below we read only one value, the contents of cell 2D, but you can read complete ranges of cells both rows and columns. Remember that the first cell in Python is indexed as (0,0).

```python
import xlrd

#   Open xls file and go to sheet 0
book = xlrd.open_workbook("data.xlsx")
first_sheet = book.sheet_by_index(0)

#   Extract value in cell 2D
valor = first_sheet.cell(1,3).value
```

Read a *csv* file or an Excel *xlsx* file with Pandas

The Pandas module allows you to read and manage files from a variety of sources, and it is a great tool for Data Science. Information is stored in a dataframe object.

```python
import pandas as pd

# Read an Excel file

my_dataframe = pd.read_excel("Data_prod.xlsx",
                              "Sheet2")


# Read a csv file

my_dataframe1 = pd.read_csv("Data_mining", sep=";",
                             header = 0)
```

179

Read a *pdf* file

You can read also a pdf (Portable Document Format). PDFs incorporate hidden information, which are formatting and pagination characters, and embedded graphics. For this reason, strange characters can be expected, in addition to the text itself.

```python
import PyPDF2

pdfFileObject = open("C:\trees.pdf", 'rb')

pdfReader = PyPDF2.PdfFileReader(pdfFileObject)

count = pdfReader.numpages

# Create a list with a string per page

list = []

for i in range(count):
    a = pdfReader.getPage(i).extractText()
    list.append(a)

pdfFileObject.close()
```

Writing in text files

The Python sentence 'write', is used in order to write strings of alpha-numeric characters. If your data are float or integer numbers, you are ok. However, if they are combinations of numbers and characters, first you need to convert them into strings with 'str' sentence. The example below illustrates the different posibilities.

Note that, we can also add information to the end of a file by specificyng the option 'append' when opening the file.

```
arch1 = open("newfile1.txt", "w") # w - write, r -
    read, a - append

arch1.write("This is the first line\n")
arch1.write("and this is the second line.\n")
arch1.write('3.14\n')
arch1.write(str(mat))
arch1.close()
```

Writing to Excel files

Writing in an Excel file can be done from cell to cell. In addition, you can perform algebraic operations between cells and write the result in an additional cell.

```
import xlwt

wb = xlwt.Workbook()
ws = wb.add_sheet('Sheet1')

# Row and column

ws.write(0,0,"Tree")
ws.write(0,1,"Leaves")
ws.write(0,2,"Type")

ws.write(1,0,"Oak")
ws.write(1,1,"big")
ws.write(1,2,2)

ws.write(2,0,"Olive")
ws.write(2,1,"small")
ws.write(2,2,6)

ws.write(3,2, xlwt.Formula("C2+C3"))
wb.save('example.xls')
```

181

Download / upload files from a remote server

In many cases it is necessary to import or download files from a remote server, and also the reverse operation.

```python
# Retrive a file

from ftplib import FTP

# Remote access

ftp = FTP(remote_server, user, password)

# Navigate to desired directory

ftp.cwd(directory_name)

# Get a directory listing

ftp.retrlines("LIST")

# Remote file we want to retrieve

fich_remoto = "file"

# Local name
fich_local = "fule.txt"

# Retrieval

mylocalfile = open(fich_local, 'wb')
ftp.retrbinary('RETR ' + fich_remoto,
                mylocalfile.write, 1024)
ftp.quit()
mylocalfile.close()


# Upload a file

import ftplib
```

```python
session = ftplib.FTP('example.com',
                     'username','password')
file = open('cup.mp4','rb')
session.storbinary('STOR '+'cup.mp4', file)
file.close()
session.quit()
```

It is not the mountain we conquer, but ourselves (E. Hillary)

Appendix B

Packages Contents

Raw Python has a limited capability for Engineering and Science. However, there are great packages or libraries that extend the possibilities to almost any technical field. Very often, almost every week, the Python Users Community improves the components of these libraries or gives birth to new tools. In addition, the concept of programming objects and platforms for the development of operating systems has not been discussed in this book, but it is a fascinating area in continuous expansion.

The most important libraries from the scientific computing viewpoint are summarized in this Appendix.

Scipy

SciPy is an open source library of mathematical functions. The power of Python is greatly expanded with this suite. Some of the areas with dedicated routines are the following ones:
https://www.scipy.org/

- cluster - Clustering algorithms

- constants Physical and mathematical constants

- fftpack - Fast Fourier Transform routines

- terpolate - Interpolation and smoothing splines

- io - Input and Output

- linalg - Linear algebra

- maxentropy - Maximum entropy methods

- ndimage - N-dimensional image processing

- odr - Orthogonal distance regression

- optimize - Optimization and root-finding

- signal - Signal processing

- sparse - Sparse matrices and associated routines

- spatial - Spatial data structures and algorithms

- special - Special functions

- stats - Statistical distributions and functions

- weave - C/C++ integration

Numpy

NumPy is the fundamental package for scientific computing with Python. Some of the most interesting features are listed next.
http://www.numpy.org/

- Array creation routines

- Array manipulation routines

- Binary operations

- String operations

- C-Types Foreign Function Interface

- Datetime Support Functions

- Data type routines

- Linear algebra

- Mathematical functions with automatic domain

- Discrete Fourier Transform

- Financial functions

- Input and output

- Linear algebra (numpy.linalg)

- Logic functions

- Truth value testing

- Array contents

- Logical operations

- Mathematical functions

- Matrix library (numpy.matlib)

- Polynomials

- Random sampling

- Statistics

Other Packages

- **Matplotlib** - It is a versatile library to generate graphs and plots from the many different sources of data generated in Python.
  https://matplotlib.org/

- **Pandas** - This library is an extension to NumPy in order to handle data from a variety of sources, and to perform statistical analysis.
  http://pandas.pydata.org/

- **Sympy** - A library to expand the Python universe with symbolic handling of algebraic operations.
  http://www.sympy.org/en/index.html

- **Mayavi** - Easy and interactive visualization of 3D data.
  https://pypi.org/project/mayavi/

- **Random** - Generate pseudo-random numbers.
  https://github.com/python/cpython/tree/3.7/
  Lib/random.py

- **Openpyxl** - Read and write Excel files.
  https://openpyxl.readthedocs.io/en/stable/

- **Seaborn** - Visualize statistical data.
  https://seaborn.pydata.org/

- **Bokeh** - Web visualizing
  https://bokeh.pydata.org/

- **Tkinter** - Interactive screens.
  https://docs.python.org/2/library/tkinter.html

- **Py2exe** - Python scripts compiler.
  www.py2exe.org/

Appendix C

Some files for you!

In this website you will find some of the files used in this book, just enter the url or scan the QR code.

```
https://drive.google.com/drive/folders/1b7AlMPTQMxQ6J8M
 Qa_3xgXQLO6Fx9uR3?usp=sharing
```

Solve it with Python!

Be good!

www.ingramcontent.com/pod-product-compliance
Lightning Source LLC
LaVergne TN
LVHW052100060326
832903LV00060B/2343